THE GENESIS OF
SCIENCE

THE GENESIS OF
SCIENCE

THE STORY OF GREEK IMAGINATION

STEPHEN BERTMAN

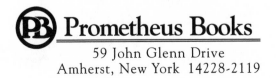

Prometheus Books

59 John Glenn Drive
Amherst, New York 14228-2119

Published 2010 by Prometheus Books

Inquiries should be addressed to
Prometheus Books
59 John Glenn Drive
Amherst, New York 14228–2119
VOICE: 716–691–0133
FAX: 716–691–0137
WWW.PROMETHEUSBOOKS.COM

14 13 12 11 10   5 4 3 2 1

Library of Congress Cataloging-in-Publication Data

Bertman, Stephen
    The genesis of science : the story of Greek imagination / by Stephen Bertman.
        p. cm.
    Includes bibliographical references and index.
    ISBN 978–1–61614–217–9 (cloth : alk. paper)
    1. Science, Ancient. 2. Science—Greece—History. 3. Science—History. 4. Science, Ancient. I. Title.

Q127.G7 B47 2010
509.38—dc22

2010022914

Printed in the United States of America on acid-free paper

For Elaine,
who lit a thousand candles

BROTHERHOOD

Homage to Claudius Ptolemy

I am a man: little do I last
and the night is enormous.
But I look up:
the stars write.
Unknowing I understand:
I too am written,
and at this very moment
someone spells me out.

Octavio Paz

Translated by Eliot Weinberger, from *Octavio Paz, Collected Poems 1957—1987*, copyright © 1986 by Octavio Paz and Eliot Weinberger. Reprinted by permission of New Directions Publishing Corp. and Gerald Pollinger, Ltd.

CONTENTS

PART III. FROM ANCIENT GREECE TO THE MODERN WORLD

PART IV. ANCIENT SCIENCE BEYOND THE MEDITERRANEAN

ACKNOWLEDGMENTS

This book was inspired by a simple question: "Who invented science?" The answer, as I soon learned, was the Greeks. But like all good questions, that question provoked another, more profound one: "Why *them*? What was so special about the Greeks that made *them* and not some other people the world's first scientists?"

Like many a writer, I assumed that once I found the answer, publishers would be eager to share my discovery with their readers, especially in an era like ours, in which the discoveries of science play such a far-reaching role in everybody's life. But I was wrong: finding a publisher proved a much more daunting task than I ever would have imagined. As it turned out, most commercial publishers had little faith in the intellectual curiosity of today's book buyers, while most academic publishers felt it was beneath their dignity to publish a book written for a wide audience.

To his credit, my trusty literary agent, Ed Knappman, never lost faith in the notion that there are enough intelligent readers out there still interested in questions of enduring significance, and enough editors out there brave enough to publish books on uncommon but deserving themes.

With poetic justice, the publisher who came to our rescue was none other than Prometheus Books, a house named for the very Greek god who once saved humanity by bringing people the warming gift of fire and endowing them with the arts and sciences that energized civilization. My deepest gratitude, therefore, goes to Prometheus's perceptive executive editor, Linda G. Regan, and its courageous editor in chief, Steven L. Mitchell.

There are others without whom this project would have been impossible to complete. I owe immense thanks to those anonymous staff members in countless libraries across Michigan who helped me obtain out-of-print books for my research, especially the long-suffering but gracious staff members of my hometown library in West Bloomfield, whose drive-up window became for me a kind of second home. I am also genuinely indebted to Jennifer Belt and Liam Schaefer of Art Resource, New York; Jose Fernandez of Photolibrary, New York; Christine Crosbie and Valerie Hatten of the Ontario Science Centre; Tom Vine of London's Science Museum; and Dr. Will Noel of the Walters Art Museum for their ever-patient and unfailingly personal service in securing the beautiful photographs that now illuminate and animate my text.

11

Special thanks go to Drs. Paul Keyser and Georgia Irby-Massie for their careful reading of my original manuscript, and to Julia DeGraf for her copy-editing skills.

Last, but most important, I must thank my precious partner and wife, Elaine, whose persistent Friday evening prayers that "Steve's book get published" somehow reached their heavenly destination.

PROLOGUE

THE ESSENCE OF SCIENCE

"Sit down before fact as a little child, be prepared to give up every preconceived notion, follow humbly wherever and to whatever abyss nature leads, or you shall learn nothing."
—Thomas Huxley

At its finest, the pursuit of science can be profoundly rewarding because the discovery of ultimate truth is an exhilarating, even mystical, achievement unmatched by any other—one that can benefit humanity as perhaps nothing else can.

We must then ask: What are the essential qualities that define science?

Prominent among them is an uncompromising humanism and an unadulterated rationalism. Humanism holds that people have the right to investigate the nature of things to determine their root cause, a right uninhibited by social convention and unrestrained by religious dogma. Rationalism holds that the means to find that truth is already in humanity's possession, for it is not revelation but reason, an innate power that can illuminate the dark corners of humankind's own ignorance, prejudice, and superstition. The inevitable enemy of science is an authoritarianism—whether human or what some attribute to be divine—that interposes its arbitrary will between the questioning mind and the object of its desire.

The force that energizes rationalism is curiosity, a childlike inclination to ask questions—especially "Why?"—an inquisitiveness that for the scientist must be guided by a mature and resolute refusal to be satisfied by superficial or convenient answers, coupled with an abiding determination to persevere patiently in the search for truth. As a consequence, the life of the scientist may be marked by a heroic individualism that, though at times wounded, is undefeated by disappointment, frustration, and failure. The scientist must also embody a tireless capacity for self-criticism lest he or she mistake pride for certainty. And because the aims of scientific inquiry must very often be stated in quantitative terms, the investigator must always be dedicated to precision and exactitude.

Together with its willing partner technology, science has been the most powerful force to shape the modern world. Yet the roots of science go back to ancient times.

The true inventors of science are echoed in the names of its branches. From "physics" to "biology," from "mechanics" to "psychology," the vocabulary of science is derived from Greek. But the Greeks did not merely bequeath us their words. We also inherited from them an inquiring mind, a mind that hungered to understand the natural world and to find humankind's place within it.

As we will come to see, science took root in ancient Greece because the seminal principles of Greek civilization—humanism, rationalism, curiosity, individualism, the pursuit of excellence, and the love of freedom—were uniquely compatible with science's own essence.

PART I:

☙☙☙☙☙☙☙☙☙

THE ORIGINS OF GREEK SCIENCE

Figure 1: Portrait of the hero Odysseus from the Grotto of Tiberius at Sperlonga, Italy. Credit: Erich Lessing/ Art Resource, NY.

Chapter 1

THE GENIUS OF GREEK CIVILIZATION

. . . that which we are, we are,
One equal temper of heroic hearts
Made weak by time and fate, but strong in will
To strive, to seek, to find, and not to yield."

—Tennyson, *Ulysses*

Why the Greeks? Why was science—the methodical investigation of the physical universe and the living beings within it—a *Greek* invention and not the creation of some other civilization?

It is a question of profound significance and it goes to the very heart of this book. For if we satisfy ourselves with making a mere compilation of separate discoveries rather than seeking the unifying genius behind them, we will have settled for the "what" or "how" of history and never touched the "why."

French scientist Louis Pasteur once declared: "Where observation is concerned, chance favors only the prepared mind."[1] Pasteur's point was that only those who are mentally ready will recognize the significance of what lies on the ground before them. Others can walk by and never notice, or—worse—never care, preoccupied as they are with mundane concerns or blind to the possibilities that lie undisclosed beneath the surface of what most might call reality. Therefore, truth, like beauty, is in the eye of the beholder—not because it is subjective, but because it can so easily be overlooked. Unlike others who came before them, the ancient Greeks recognized that truth for the first time and thus more passionately and more critically observed the world around them.

What was it that made the Greeks this way, that allowed them to see—and impelled them to search for—what millions of others had missed for millennia?

Surely the land of Greece itself was an unlikely mother of geniuses. Its

17

soil was rocky and poor, lacking the fertility and abundant water that had long before blessed the river valleys of the Near East where civilization itself had been born. Scattered on islands or separated from each other by mountains, the Greek people would seem to have been too fragmented to ever have left a collective and decisive mark on history. The miracle is that they did, not by drawing upon the resources around them but by maximizing the potential that lay within. From a critical examination of their surviving literature and art, we can reconstruct the unique traits of personality that made the ancient Greeks who they were, and from these ethnic characteristics we can trace the genesis of the scientific attitude that has come to define the modern world.

First among these traits was rationalism, a reliance on the intellect to find answers and solve problems. Rather than prayerfully turning outward for help, the Greeks instead persistently turned inward and used their brains, convinced that the mind was a fit instrument to accomplish any task. While not underestimating the capricious power of the gods, they looked upon the universe, however complex, as a fundamental expression of order accessible to the mind. If the universe was a lock, then intelligence was the key that could open it. Their leaders in this enterprise were philosophers, not "*wise* men" but "*lovers* of wisdom," as the word's etymology shows.[2] Indeed, ancient Greece produced the world's first philosophers, men aggressively dedicated to the rational pursuit of truth.

It was their rationalistic "ear" that enabled the Greeks for the very first time to hear the language of the cosmos. That language was mathematics— a mathematics beyond the mere measurement or calculation of particular things but dependent upon immutable laws to which all the phenomena of this world were, are, and forever will be, obedient. In learning this universal language of space and time, the Greeks became conversant with nature's most fundamental relationships.

Though not averse to applying their knowledge to the everyday world, Greek thinkers more often than not preferred the development of theories to their practical application, enamored as they were with reason's abstract purity.

The rationalism of the Greeks was combined with humanism, a pride and confidence in their own human potential. This pride was in part born of their successful struggle to survive, but in larger part it was engendered by the firm conviction that humanity can possess an experience no god could ever have: the thrill of victory won by risking everything against great odds. Such a thrill was reserved for mortals precisely because their lives were inherently so fragile and their powers so finite. It was a thrill the Olympian gods, omnipotent and immortal, could never share, for because man had to die, only man could most fully live. Thus, confronted with the choice of becoming a god or remaining human, the hero Odysseus (fig. 1, above) chose to retain his humanity, even though it meant he had to eventually relinquish his life.

The Greek zest for living was driven by a compulsion to excel. Knowing they could not exist for all time, they determined to achieve a different kind of immortality by performing deeds that would ensure their being remembered forever. Thus the hero Achilles was willing to die young on the battlefield of Troy as long as his name remained indelibly impressed on the minds of future generations. In addition to pursuing excellence on the battlefield or, more peacefully, in the Olympic Games, the Greeks also pursued it by creating enduring works of literature and art, and by reaching new heights in the search for knowledge, never retreating in debate from the battleground of ideas. Their pursuit of excellence was also marked by a passion for perfection and a dedication to detail that would eventually contribute to the precision essential both to the fine arts and to science.

Another of their distinguishing traits was restless curiosity, an insatiable hunger to understand themselves and their world. This trait would not only prove instrumental in making the Greeks the world's first scientists, but it also explains their achievements in other fields. Drama, for example, itself a Greek invention, represented the quest to understand the motivations for and the consequences of individual human behavior. The field of history, another Greek innovation, examined the implications of that behavior on a wider stage. Indeed, the very word *history* in Greek means "research." Even democracy, yet another Greek invention, was compatible with objective science, for as critic Mary McCarthy once observed, "In science, all facts, no matter how trivial or banal, enjoy democratic equality."[3]

Underpinning all of these national characteristics was a fierce love of freedom and of individualism. Without freedom of thought and unfettered imagination, science could not exist. And without heroic individualism, the bastions of stultifying convention could never be breached.

All these distinctive national characteristics would set the Greeks apart from the earlier civilizations of the ancient world.

Chapter 2

SCIENCE BEFORE THE GREEKS

PREHISTORY

The Stone Age is by far the longest chapter in human history, but one that speaks to us wordlessly because of the absence of writing. For *pre*history, the time before historical records, our only witnesses are mute stone tools, fossilized skeletons, and captionless works of art.

This prehistoric period began as far back as two and a half million years ago with the appearance of the first human-like creatures on our planet and culminated with the emergence of *Homo sapiens* around 100,000 BCE. Yet even in this primordial era, there is persuasive evidence that scientific methods were practiced.

The fundamental focus of prehistoric men and women was survival. Because their search for food depended on hunting, fishing, and gathering, their very existence necessitated careful and patient observation (of the habitats, behavior, life cycles, and migration patterns of animals) and the meticulous collection of empirical data (about which plants were nutritious or potentially curative, and which were harmful). Without the help of writing, all knowledge had to be remembered and deliberately passed on by word of mouth from one generation to the next.

In an Ice Age environment, the manufacture of artificial warmth became critical to human survival. After discovering fire by chance and cautiously learning its virtues (generating heat, warding off predators, and making food more tasty and digestible), primitive humans sought the methods by which they could re-create it at will and sustain its flame.

Through trial and error, humans acquired the means to make tools, striking one stone against another or applying sufficient pressure to produce sharp edges

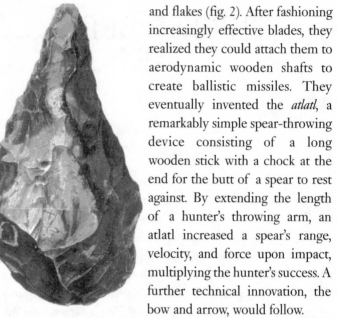

Figure 2: Flint implement from the Paleolithic, or Old Stone, Age. From Ernest A. Parkyn, An Introduction to the Study of Prehistoric Art (New York: Longmans, Green and Co., 1915).

and flakes (fig. 2). After fashioning increasingly effective blades, they realized they could attach them to aerodynamic wooden shafts to create ballistic missiles. They eventually invented the *atlatl*, a remarkably simple spear-throwing device consisting of a long wooden stick with a chock at the end for the butt of a spear to rest against. By extending the length of a hunter's throwing arm, an atlatl increased a spear's range, velocity, and force upon impact, multiplying the hunter's success. A further technical innovation, the bow and arrow, would follow.

In confronting the power of brute nature, prehistoric artists inspired by fear and awe painted the walls of their caves with portraits of the fleet and powerful beasts that were their prey. These paintings made by our earliest *Homo sapiens* ancestors reveal a keen eye for anatomical detail. The chemical pigments for painting were taken from the ground: yellow, red, and brown from ochre (iron oxide) found in soil; violet from manganese ore; and black from the charcoal of campfires. They were all blended together with animal fat or albumen and applied with fingers, feathers, and twigs or sprayed onto a wall's surface by blowing through a hollow stalk. Except for an occasional handprint left as a signature on a wall, Paleolithic hunters portrayed themselves only rarely, perhaps because they felt so small in the face of a hostile environment. Both the startling reality of death and the bloodstained mystery of birth fascinated the primitive mind and evoked a spiritual response in sculpture that celebrated the fecundity of the female in repopulating the tribe.

Eventually by about 10,000 BCE in various parts of the world, a "Neolithic Revolution" took place as people discovered how to produce food by growing grain and domesticating animals. As a result, farming settlements grew up, populations increased, and new skills of survival and adaptation emerged. By the third millennium BCE, cultures across Western Europe, most notably in Britain at Stonehenge, were arranging colossal stone blocks in monumental circles to mark and predict solstices and to celebrate the changing seasons of the agricultural year.

ANCIENT EGYPT

In rare parts of the world between 3500 and 3000 BCE, the seeds of the Neolithic Revolution fell upon soil so fertile and well watered that huge populations flourished due to a superabundance of produce and livestock. One of these rare places was Egypt.

An ancient Greek traveler named Herodotus once described ancient Egypt as "the gift of the Nile."[1] Herodotus meant that, were it not for the river's blessings, there never would have been an Egypt. True enough, for the river not only provided ample water but also a regular renewal of the land's fertility, turning what would have been barren desert into an agricultural paradise.

Every year in July when the star Sirius appeared on the horizon at dawn, the waters of the Nile slowly began to rise, fed by melting snows thousands of miles to the south. By August, the waters of the Nile had crested and spilled over the river's banks, inundating farmers' fields. When the waters eventually ebbed in September, they left behind a fresh layer of silt that replenished the fields' fertility. Egypt's astronomers, looking to the east, closely watched the stars to predict the impending flood. Upriver, with the aid of a vertical depth-gauge, other priests diligently measured the Nile's rising swell, for the river was seen not just as a fixture of nature but also as a benevolent god whose coming was awaited by the people with joyous anticipation.

By digging reservoirs and irrigation canals, farmers could water the new crops that would spring up. To raise the water up from the river and pour it into a canal, the ancient Egyptians invented a labor-saving mechanical device known in Arabic as a *shaduf*, a device still employed by the peasants of Egypt today (fig. 3). Based on the principle of the lever, a shaduf is a long pole mounted like a seesaw on an elevated post that acts as a fulcrum. At one end of the long pole is a basket, and at the other end, a counterweight. After the farmer pulls the basket down and dips it into the river to fill with water, the counterweight helps him to lift it back up. Swinging the pole around, he then empties the water-laden basket into an irrigation ditch, and repeats the process over and over again.

Figure 3: Egyptian peasants using shadufs *to raise water from the Nile. From Adolf Erman,* Life in Ancient Egypt, *trans. H. M. Tirard (London and New York: Macmillan, 1894).*

Because the Nile flows from south to north, and the prevailing winds in Egypt blow from north to south, transportation was easy. Riverine commerce was so effortless that the ancient Egyptians were inspired to invent the world's earliest sailboats that, with their sails down, floated northward toward the Mediterranean and, with their sails raised again, glided back to the south. In addition to aiding transportation, the Nile unified the communities along its length geographically, forming the world's first nation-state, a country protected from invaders by forbidding deserts to the west and east.

The seasonal regularity of the Nile and the changeless and protective character of the deserts infused Egyptian consciousness with a sense of permanence and dependability. Besides river and desert, the Egyptians' sense of security was reinforced by the sun, whose life-giving radiance by day never failed to shine. The visibility and spiritual significance of the sun would inspire Egyptian astronomer-priests to devise the world's first solar calendar. In their mind's eye, they envisioned the orb of the sun traveling across the sky while being borne on a divine boat or being rolled across the sky by a gigantic but invisible sacred scarab beetle.

Though the sun-god Ra might "die" in the west, he was reborn again in the east in a never-ending, life-affirming cycle that held out to the ancient Egyptians an intimation of their own immortality. Perhaps they inferred that, like the sun, we, too, do not truly die but can somehow be reborn in another dimension.

Ironically, this metaphysical inference was supported by the seemingly lifeless desert, for corpses laid in its hot shallow sands retained their distinctive features for centuries after their interment due to the desert's ability to dehydrate and thereby preserve their flesh. If the body does not perish, the Egyptians optimistically reasoned, perhaps the human spirit whose home is the body is similarly imperishable. Consequently, they buried their loved ones with personal possessions, convinced they could enjoy them in the afterlife.

In any civilization, however religious, piety can coexist with greed. Therefore, even as funeral ceremonies were being concluded, thieves were already hiding behind nearby sand dunes waiting for the chance to dig down and rob the graves of their treasure. Upon the disheartening discovery that such robberies had been committed, the families of the dead reburied their loved ones in a way that would make their happy afterlife more secure. Consequently, graves were dug deeper to foil the thieves.

But greed is a powerful incentive, and even these deeper graves were often robbed soon after, or sometimes long after, the final farewells had been uttered. When the relatives of the deceased returned to the grave site to undo the damage done by the thieves, they were even more horrified than before, for, as they looked down into the grave, they saw that the flesh of their loved one was no longer intact but had decomposed, imperiling the integrity of the vessel the soul needed for its spiritual journey. While a shallow grave permitted the heat of the sun to preserve flesh, a deeper grave, rather than giving a body greater protection, had subjected it to the decay-inducing moisture that lurked in the ground beneath the sands.

Not knowing what to do, the ancient Egyptians finally solved their dilemma through the science of chemistry. Just as Egypt was rich in sand, so was it rich in a variety of chemical compounds that had preservative properties. Egyptologists collectively call these chemicals *natron*, though there were a number of different compounds the ancient Egyptians used: sodium chloride (ordinary salt), sodium carbonate, sodium bicarbonate, and sodium sulphate. Applied by ancient morticians, natron had three main positive effects. First, it was a powerful drying agent that, by eliminating moisture in the flesh, inhibited the activity of decay-producing bacteria. Second, it was a powerful degreaser, dissolving moisture-laden fat cells that could similarly promote decay. And third, it was a powerful antimicrobial disinfectant. The Egyptians effectively theorized that natron might preserve human flesh, perhaps from their experience in using salt to preserve meat.

Protecting a corpse from decay in a hot climate, however, did not simply mean treating it with natron. Though natron was the key ingredient in engineering eternal life, the Egyptians followed an elaborate procedure both before and after the natron was applied. The techniques of what is known as mummification were closely guarded ritual secrets, but our Greek friend Herodotus was able to learn some of them from the priests he fraternized with.[2] Altogether, these secrets constituted a science of immortality. Herodotus describes three methods of embalming; the choice ultimately determined by the wealth of the deceased.

The most elaborate method involved an autopsy of sorts, in which all organs and entrails were surgically removed from the body to reduce the wetness of the corpse. Paradoxically, it might seem to us, the brain was taken out and disposed of. The Egyptians did not understand its mental function, believing instead that the organ of thought and feeling was the human heart, which was dutifully left inside the corpse. Other organs—liver, lungs, stomach, and intestines—were removed in order to be treated with natron, after which they were either stored in funerary jars or reinserted into the body cavity. Next, the body—with soft organs removed—was packed with natron and set out in the open air under the sun beneath a heap of natron crystals in order to eliminate all residual moisture. After about thirty-five days, the corpse was cleaned off and stuffed with rags or sawdust to fill it out and make it look more lifelike. Then, after the abdominal incision was sutured, the skin was anointed with imported cedar oil to keep it supple. The entire body was then intricately bandaged in multiple layers of linen, and coated with a layer of molten resin (known in Arabic as *mummia*, from which the word "mummy" comes) to serve as a final barrier against decay (fig. 4). This step is not unlike the rust-inhibiting undercoating applied to a new car. After the mummy was put inside a protective and elaborately painted case, various prayers and incantations were recited prior to its placement in its stone sarcophagus and tomb.

Because ancient Egyptian religion made the body the final resting place of the soul and because preservation required such an elaborate system of dissection and treatment, the Egyptians acquired more hands-on knowledge of the human body's internal parts and structure than did any other ancient civilization. However, as we have noted in connection with the disposal of the brain, the precise function of a particular organ often eluded them. Nevertheless, because of their surgical experience and their expertise in pharmaceuticals, the Egyptians, before the coming of the Greeks, were regarded by the cultures of the ancient Mediterranean as the masters of medical science.

Figure 4: The unwrapped mummified head of Pharaoh Ramses II. From P. S. P. Handcock, The Latest Light on Bible Lands, *2nd ed. rev. (London: Society for Promoting Christian Knowledge, 1914).*

Our knowledge of Egyptian medicine comes from their ancient textbooks and from the Egyptian dead themselves. Though mummies are today housed in museums, mummies are themselves museums of boundless biological information. They have become one of the main subjects of paleopathology—a contemporary science that explores the topic of disease in antiquity—primarily through the use of autopsies and tissue studies. From their research on Egyptian mummies, investigators have uncovered the presence of multiple maladies: hardening of the arteries; evidence in the kidneys of high blood pressure; scars left by heart attacks; punctured eardrums; abscessed teeth; respiratory conditions such as bronchitis, pneumonia, anthracosis (caused by the inhalation of smoke in poorly ventilated areas), and silicosis (brought about by the inhalation of sand); severe appendicitis; gallstones; kidney stones; inguinal hernia; rectal prolapse; gout; intestinal parasites; bedsores; arthritis, slipped discs; fractures and rheumatism of the spine; ovarian tumors; and bone cancer. Significantly, most ancient prescriptions found in Egypt were for infections of the eye, a widespread problem in Egypt even today because of the persistence of similar pathogens in the environment.

Dental problems also took their toll, though their cause, surprisingly, was not due to cavities. For most of the Egyptians' history, their diet was relatively low in sweet, sticky foods. Most dental problems were the result of attrition, the wearing away of the tooth's protective enamel surface. The chief cause seems to have been windblown sand that worked its way into the food supply and literally ground the Egyptians' teeth away as they chewed.

The healers of ancient Egypt certainly had their work cut out for them.

Their approach was twofold. For conditions that had an obvious origin—blows to the head, for example, or other types of wounds—the treatment was practical and methodical. For conditions in which the causes were *not* apparent but instead might be due to some mysterious agent—a dead person's curse, perhaps, or the anger of a god—the treatment was spiritual: magical spells and incantations were recited in an attempt to placate the offended spirit, or unpleasant-tasting potions would be administered to drive it out. In some cases in which a practical approach was taken, spells might also be recited for added insurance. In other cases, patients might be sent to a "continuing care" facility attached to a temple, where the power of faith combined with proper rest, it was hoped, would restore their health.

The most convincing evidence of the Egyptians' scientific approach comes from their medical textbooks. Written on papyrus, they are the earliest such treatises in history to survive and contain the world's oldest medical terminology. These papyrus scrolls, which most likely once belonged to practicing physicians (*senu*), date back to the days of Egypt's New Kingdom (approximately the time of King Tut). But some scrolls incorporate knowledge that goes even further back to the Pyramid Age, a thousand years or more before Tut's birth. Egyptian tradition, in fact, declares that Djer, a pharaoh of the First Dynasty, authored a book on human anatomy. Imhotep, moreover, the architect who designed Egypt's very first pyramid, was revered as a god and also esteemed as the father of medicine.

The most scientifically impressive of these textbooks is the *Edwin Smith Medical Papyrus*, named for its nineteenth-century American owner and now in the collection of the New York Academy of Medicine. Its subject is surgery, and its approach is highly rational and systematic. It presumes an ancient form of triage: every condition is categorized as either treatable by surgical methods, possibly treatable, or beyond treatment. After the condition has been named and categorized, its symptoms are described, its cause is diagnosed, and a specific treatment plan (if possible) is recommended. Elsewhere, tomb paintings show a surgeon applying a splint and another setting a dislocated shoulder. Some surgeons, we learn, even practiced cosmetic surgery to repair sagging breasts.

From documents like the *Edwin Smith Medical Papyrus*, we know that the Egyptian medical profession numbered specialists among its ranks: not only surgeons, but specialists in ophthalmology, podiatry, dermatology, gastroenterology, gynecology, and dentistry. However, many of the medical diagnoses were misguided, and the treatments recommended were useless (a criticism, we must quickly add, to which even modern medicine is not completely immune). Despite their familiarity with cadavers, the greatest shortcoming of Egyptian physicians was their ignorance of how a living human body functions. Though they had identified the pulse as "the voice of the heart," they did not comprehend the true nature of the circulatory system, and regarded arteries, veins, muscles, sinews, nerves, and ducts as essentially similar "vessels" belonging to the same system. Furthermore, apart from noting that an injury to the head could impair motor function, they did not appreciate the neurological and intellectual significance of the brain.

Judging by the mouths of mummies so far examined, Egyptian dentistry seems to have been severely limited. Findings include a single instance of a drained abscess and a weak tooth that was bound to a stable one with gold wire.

When it came to pharmaceuticals, the Egyptians were creative, drawing upon a wide variety of animal, vegetable, and mineral substances, which they often combined in the scattershot hope that one might work where others failed. For oral consumption, the *materia medica* were diluted in water, milk, beer, or wine and sweetened with honey. For topical application, the same substances were blended with honey or animal fat. Alas, because the precise meanings of many of the items in the prescriptions are hieroglyphically obscure, we are at a loss to know exactly what the active ingredients, especially the herbs, actually were. There are some notable exceptions: a mixture of honey, cream, and milk was used to treat a sore throat and cough; a penicillin-like fungus that grew on stagnant water was applied to open wounds and sores; and, where all other treatments had failed, a child might be induced to swallow a whole skinned mouse! (This is sadly confirmed by the mummified remains of ancient Egyptian children that still contain the shriveled bodies of the mice in their innards.) The scrolls also list formulas for removing blemishes from skin, a recipe for preventing hair from falling out, a variety of recommended aphrodisiacs, and even a spell for making a rival go bald. Though ignorant of the germ theory of disease, Egyptian doctors nevertheless concluded that household pests might somehow spread infection, and so they offered remedies for ridding one's house of vermin, fleas, and flies.

The elaborate preparation of a corpse (described above) was more than matched by the preparation of the structure meant to house it for all eternity and guarantee it everlasting life. The most elaborate of these structures were the pyramid-shaped sepulchers intended for Egypt's kings. Just as mummification indirectly promoted biological and chemical research, so the design and construction of pyramids worked hand in hand with developments in mathematics and astronomy.

The pyramids—the royal tombs of Egypt—have been rightly described as man-made mountains built with gem-cutting precision. Separated from our own time by almost five thousand years, they are Egypt's most monumental expression of the longing for eternity and permanence. Of all the theories that have been advanced to explain their distinctive triangular shape, perhaps the simplest and most persuasive is that they attempted to architecturally replicate the spreading rays of the sun as it breaks through a cloud, for the sun was Egypt's most visible symbol of rebirth.

Of all the pyramids, the largest was built at Giza by Cheops, or Khufu, around 2600 BCE. Forty stories tall, it is the only one of the Seven Wonders of the Ancient World that still stands. Constructed of over two million blocks of limestone weighing between two and a half and fifteen tons, the Great Pyramid is remarkable for the painstaking precision with which its outer blocks were cut and finished—so accurately that a credit card cannot be slipped between them. Equally remarkable is the fact that the Egyptians had no pulleys or wheeled vehicles to move rock and possessed only simple tools made of wood, stone, or copper. What they lacked in equipment, however, they made up in manual labor, systematic organization, and sheer patience.

Painstaking precision, that hallmark of the sciences, is equally evident in the Pyramid's siting. Laid out by surveyors with an east-west orientation to invoke the promise of immortality proffered by the sun, the sides of its 756-foot-by-756-foot base are only seven inches short of forming a perfect square, while its north-south axis misses true north by only one-tenth of a degree.

Because of the prominence of the heavens in Egyptian thought, the pyramids of Giza were aligned with the stars. By identifying circumpolar stars with the naked eye or by tracking the course of a single star across the sky, astronomer-priests were able to find true north and use it to mark off a planned pyramid's axes. Narrow shafts hidden within the Great Pyramid pointed to the ancient North Star (Alpha Draconis), providing the soul of the dead pharaoh with a trajectory it could use for its journey into space, while another internal spirit-shaft pointed south toward the constellation Orion, long identified by the Egyptians with Osiris, their god of resurrection.

Unlike the later abstract mathematics of the Greeks, Egyptian mathematics was elementary and doggedly practical, having arisen as a ten-finger decimal system in response to the need to measure a plot of land, determine the amount of bread needed to feed an army, or calculate the number of bricks required to build a wall. Problems in multiplication and division were done by ponderously adding or subtracting sums, and, except for a symbol for two-thirds, only fractions with a numerator of one—one-quarter, for example, or one-half—were used. Even so, Egyptian scribes were capable of solving simple problems in algebra and plane and solid geometry, including calculating area, volume, and slope (so critical to the erection of a pyramid). They even discovered the approximate value of pi two thousand years before the Greeks. With supreme dedication and by applying the profoundly simple methods they already understood, the ancient Egyptians rose from the Nile mud to reach for the stars.

ANCIENT MESOPOTAMIA

About the time civilization was born in Egypt (around 3500 BCE), it also arose in ancient Iraq.[3] From the days when Greek travelers first visited, ancient Iraq was called Mesopotamia, a Greek word meaning "the land between the rivers." The rivers were the Tigris (to the east) and the Euphrates (to the west).

Like the Nile in Egypt, the Tigris and Euphrates provided the preconditions for the development of a large population: ample fresh water and fertile alluvial soil, both of which helped agriculture to flourish. But, although both river-valley systems shared nature's blessings, the physical environment of the two lands was radically different. Unlike Egypt, where the annual rising of the Nile's waters was gradual and predictable, bestowing the benefits of reviv-ifying silt upon farmers' fields, the springtime flooding of the Tigris and Euphrates was sudden, violent, and unpredictable. Torrential floods could sweep away entire villages and cities built along their banks. Indeed, the bib-lical tale of Noah and the ark was inspired by an even earlier Mesopotamian legend about just such a terrifying deluge.

Two fundamental characteristics of Mesopotamia's environment—unpredictability and impermanence—instilled in its people's consciousness a pervasive mood of pessimism. Aggravating this underlying pessimism was another environmental factor, the flatness of the land, especially in the south, that invited progressive waves of invasion and imperialistic conquest.

The gods of Mesopotamia—and there were many—were endowed with personalities that reflected the natural world they ruled: they were powerful and capricious and had little compassion for humankind. At best, they could be thanked with sacrifices for their grudging blessings or placated to avoid their wrath. Unlike the biblical book of Genesis, in which man is listed as God's cosmic masterpiece, the Babylonian *Epic of Creation* makes man's cre-ation a mere footnote to the god Marduk's savage ascendancy over other deities. Likewise, when the hero Gilgamesh searches for the secret to eternal life, he is told bluntly that immortality is the exclusive prerogative of the gods and that human beings had best settle for the simple pleasures their brief lives offer.

Inhabiting a natural world that was inherently inscrutable and a humanly made world that was marked for destruction, the people of Mesopotamia hardly seemed like candidates for careers in science. Yet, as we will see, they made a number of significant contributions to scientific progress despite these disincentives.

Their main contributions were practical rather than theoretical, and some of the most notable of these involved the invention of labor-saving devices for farming. Once their cultivated fields began to spread beyond the banks of the Tigris and Euphrates, the people of Mesopotamia had to devise a way to water their crops. Recognizing that water flows downhill, they dug a series of canals that led from the rivers to their fields. However, they had to maintain the right slope over the distance required. If the slope was too steep, the water would rush through the canal and erode its bottom, deepening the channel until its water level was too low to reach thirsty crops. On the other hand, if the slope was too gradual, the canal would clog with sediment, impeding the flow of precious water.

To raise water from the rivers and feed it slowly into their canal system, the Mesopotamians employed a device still used by their descendants today, a mechanism known in Arabic as a shaduf. (As we have already noted, the shaduf was—and still is—used in Egypt, although which civilization first invented it is a matter of conjecture.)

Based upon the principle of the lever—that a shaft resting on a fulcrum can be used to multiply force—the shaduf enabled a worker to easily lift a water-filled basket, thanks to the counterweight at the end of his pole. With the shaduf, an anonymous Near Eastern genius had applied a critical principle of mechanics three thousand years before Archimedes ("Give me a place to stand and I will move the world!"). Using multiple shadufs, farmworkers could also raise water from one level to another.

To draw water up from wells, the Mesopotamians constructed chain-pumps consisting of a succession of buckets attached to a chain. As the water-laden buckets reached the top, they automatically tipped over and spilled their contents into a trough before descending again to scoop up a new load. To reduce the muscle power needed to pull the heavy buckets up, the inventor made use of another mechanical innovation of the ancient Iraqis, the pulley.

Meanwhile, by applying the principle of gravity, another enterprising Mesopotamian invented the "seeder plow," a device that saved farmers the effort of bending over to drop their seeds into the ground. As the farmer poured seeds into a funnel at the top of the plow handle, the seeds slid down a vertical channel inside the plow and swiftly dropped into the furrows sliced by the blade.

Mesopotamia, however, was not just a land of farms but a land of cities—in fact the world's first major cities—made rich and populous by the earth's bounty. As a result, Mesopotamian genius turned to the enhancement of urban life.

As early as the fourth millennium BCE, an anonymous architect realized the weight-bearing potential of an arch. Unlike the traditional post-and-lintel system used for construction from time immemorial (made up of a horizontal lintel supported by two vertical posts), the arch transferred the heavy load it carried outward and downward into the ground rather than carrying it solely on its back. This design made an arch less prone to structural failure than a stress-prone lintel. Not only that, but brick arches were easy to build—an important convenience in a land where clay was common and construction-grade timber was rare. As a bonus, an arch in a building—unlike a solid wall—required less material and allowed users to walk through its opening. Built back-to-back, such arches could also form sturdy passageways.

Long before the first brick arch was ever constructed, houses in the swampy southern part of Iraq had been fashioned by bending thick bundles of tall reeds to form tunnel-like enclosures. The inventor of the arch realized that he could substitute bricks for reeds, provided that the bricks had slanted sides to make them lean against each other. In abstract terms, he realized that a flowing curve could be replicated by a series of triangular segments that were tangent. First however, he had to figure out how to assemble the real-life components without having them fall apart. The solution was to pile up the slanted bricks higher and higher on each side of a wooden frame until a slanted keystone could be slipped in at the top to hold the complete assem-blage in place. Once the scaffolding was removed, gravity would do the rest. In effect, the construction of the first true arch was a monumental expression of applied physics.

Though the wheel had been invented in the Near East around 3500 BCE to facilitate the shaping of pottery, in Mesopotamia it was soon adapted to the purposes of transportation. Thanks to a ceremonial standard, or placard, from the city of Ur that celebrated a crucial military victory, we can still see four-wheeled chariots drawn by onagers, or wild asses, rumbling across a battlefield and crushing the bodies of the city-state's enemies. The mosaic pictures depicting this scene (plate 1), dating to the middle of the third millennium BCE, are the oldest-surviving representations of wheeled vehicles, each of which carried a driver and a spear-man. Archaeological excavations further inform us that the wheels on these mobile firing platforms were made of wood.

The invention of the chariot was based on a novel idea: attaching two wheels to a horizontal pole. Rather than using a spinning wheel's flat surface to turn domestic pottery, the chariot's rotating edge was used to accelerate the art and science of war. Thousands of years later, such chariots would become the archetypes for both the peacetime automobile and the modern military tank.

In the seventh century BCE, King Nabopolassar of Babylon built a 380-foot bridge to speed up traffic across the Euphrates (which until then had been by boat). Gigantic for its time, the bridge was supported by massive piers (twenty-eight feet wide and sixty-five feet long) that were hydrodynamically designed: the piers were rounded on the upstream side where they met the full force of the current, and were then tapered downstream like today's airplane wings. Unfortunately, the bulkiness of the piers took up half the river's width, increasing the velocity of the water flowing between them and eroding the riverbed around their foundations. Nevertheless, Babylon's great bridge managed to stand for over six centuries.

Besides their innovations in mechanics and physics, the Mesopotamians also made progress in chemistry, judging by ancient artifacts that have been unearthed: crucibles and pieces of apparatus for filtering, distilling, and extracting various chemical substances. Their scientific and technical know-how is also demonstrated by the products they manufactured, including soap, tanned leather, dyed wool, glazes, and glass, which would have required a familiarity with the properties of acids, lime, sodas, and silicates. A sophisticated understanding of fermentation was also a requisite for the manufacture of Mesopotamia's most popular drink, beer, of which over seventy varieties existed.

Perhaps the single most intriguing artifact ever unearthed in Iraq is the "Baghdad battery." Discovered in a Baghdad tomb dating to the Parthian period (about 250 BCE to 250 CE), the object consists of an oval-shaped ceramic jar containing the remains of a copper tube closed at one end, an iron rod, and some pieces of asphalt adhesive. The most convincing theory to date is that the object functioned as a primitive electric cell, or battery. If in a scale model a copper tube containing an iron rod is filled with an acid (vinegar, for example), the combination is capable of generating a half-volt of current for over two weeks. Wired together to produce sufficient current, a series of such cells could have been used to electroplate metal—such as copper jewelry—with a coating of silver or gold, a technique still employed by local craftsmen in Iraq. The tomb where the battery was discovered may in fact have belonged to one of their forebears or perhaps to the very inventor of the device, who may have been buried with it so he could continue to practice his craft in the afterlife.

The ancient Mesopotamians also applied their knowledge of chemistry to the science of medicine. Indeed, an alternate theory about the Baghdad battery proposes that its electrical stimulation may have been used to relieve pain.

When it comes to Mesopotamian pharmaceuticals, our best source of information is the royal library of the seventh-century BCE Assyrian king, Ashurbanipal. Among the 24,000 cuneiform tablets found in the ruins of his palace at Nineveh are lists of 250 vegetable substances and about 120 mineral ones cited for their medicinal properties. Mineral agents include potassium nitrate (or saltpeter), a known astringent, and sodium chloride (or salt), a recognized antiseptic. The expertise that was required to extract and purify some of these minerals further confirms the ancients' understanding of chemistry. In addition, the curative powers of many of their natural extracts (drawn from various seeds, fruits, roots, leaves, branches, barks, and gums) have been confirmed by modern science. Instructions are even given as to the best time to pick specific herbs in order to ensure their maximum effectiveness.

Texts carefully detail the medical condition to be treated, the type of medicine to be used, and the way it should be administered: orally, as a salve or ointment, as a poultice, as an inhalant, as drops for eyes or ears, or even by being blown through a hollow reed or tube into the vagina or penis. Treatments with emetics, enemas, suppositories, and hot baths are also described. Significantly, many of the recommended treatments concur with modern medical practice.

The Mesopotamian doctor would first examine his patient, noting temperature and pulse, reflexes, and the color of skin or urine. Then, following a flowchart of symptoms, he would reach a diagnosis and determine the appropriate course of treatment. Cuneiform textbooks describe the nature and treatment of intestinal problems such as obstructions, colic, and diarrhea; neurological ones such as recurrent headaches and epilepsy; rheumatism, jaundice, and gout; infestations of lice; and diseases such as tuberculosis, smallpox, typhus, bubonic plague, and venereal diseases including gonorrhea. Though the literature does not specifically discuss a germ theory of disease, an eighteenth-century BCE document warns of the dangers of contagion brought about by close personal contact. Notably, a thousand years later, Mesopotamian midwives were using a kind of gynecological "litmus paper" to test for pregnancy: a tampon soaked in a plant extract that reacted to changes in the pH value of vaginal secretions.

Because some illnesses could be caused by evil spirits (of which the Mesopotamians believed there were some six thousand!), a medical doctor, or *asu*, often worked in collaboration with a spiritual healer, or *ashipu*. The spiritual healer's job was to rid the patient of the specific demon that was causing his problem. The healer did this by helping the patient to identify what sin he might have committed that could have angered a spirit; he then proceeded to exorcise the appropriate demon. Underlying such a seemingly primitive approach to disease is the insight that physical health may be connected to spiritual health, and that an illness can be psychosomatic in nature if it is mentally induced by the pressure of worry or guilt. Mesopotamian physicians were in fact attuned to the existence of mental illness and further theorized that impotence could be of psychological origin.

One of the most fascinating aspects of ancient Mesopotamian medicine is described by the fifth-century BCE Greek traveler and historian Herodotus.[4] When a family member became ill, his relatives put him in the public square. As fellow citizens passed by, they were obligated to offer him advice if they had ever had the same ailment. In many ways, this early form of "community medicine" resembles the way people today use Internet chat rooms and forums to empathetically share experiences and discuss the latest avenues of treatment for their or their loved ones' medical conditions.

The close connection that existed between scientific medicine and spiritual healing generally parallels the relationship in Mesopotamian thought between the science of astronomy and the practice of astrology.

The people of Mesopotamia believed that their communal destinies were written in the heavens. As a consequence, whatever factual information they gleaned from their observations of the stars and planets was subordinated to the spiritual guidance that they believed this information could provide. Nevertheless, the spiritual impetus that led them to scan the skies for warnings about such things as famine or war generated substantial stores of scientific data.

Though their priests lacked telescopes and could not see the outer planets—Uranus, Neptune, and Pluto (recently demoted from planetary status)—Mesopotamians painstakingly tracked the paths of Mercury, Venus, Mars, Jupiter, and Saturn. Venus, the planet that rose earliest in the morning and set latest in the evening, was named for their goddess of love, a practice the Greeks and Romans followed in a later era. Envisioning figures among the stars in the sky, the Mesopotamians connected the stellar dots and drew pictures, giving the constellations the same names—albeit translated into Greek or Latin—that we call them today. What they saw in the night sky as a bull became Taurus; twins, Gemini; a crab, Cancer; a lion, Leo; balance scales, Libra; a scorpion, Scorpio; an archer, Sagittarius; a creature resembling a goat,

Capricorn; a man bearing water, Aquarius; and a mythical dragon, Hydra. They also divided up the sky into the now-familiar signs of the zodiac and calculated portentous horoscopes for commoners and kings. (Other priests, meanwhile, looked not to the skies but into the organs and entrails of sacrificed animals for omens of the future.) Their astronomers were also the first to identify the Milky Way. Fascinated by the changing phases of the moon, they invented a type of lunar calendar still used by pious Muslims and Jews. Indeed, the traditional names of the Hebrew months echo the very names used by Babylonian astronomers thousands of years ago. In mathematical tables and diaries they recorded the movements of the stars and planets, the cycles of lunar and solar eclipses, and the appearance of comets, accumulating an invaluable database for the later cosmic speculations of the Greeks. The spiritual need to record and predict celestial events, along with the practical need to measure farmland and count commodities on the earth, inspired the development of mathematics.

The mathematical system the Mesopotamians used was a cross between a decimal system (based on the number ten and its multiples) and a sexagesimal system (based on the number sixty). The 360-degree circle that measures space and the sixty-minute hour that governs our lives are part of Mesopotamia's enduring mathematical legacy to today's world, as is their use of a symbol for zero to make numerical calculations.

ANCIENT ISRAEL

Ancient Israel's signal contribution to religious history was monotheism, a spiritual construct that set the Jews of ancient Israel apart from their Egyptian and Mesopotamian neighbors. Beyond the history of religion, however, Hebrew monotheism had a profound influence on the development of the scientific outlook in ancient Israel, and even more so on later cultures that were ideologically conditioned by biblical thought.

Ancient Israel's unitary vision of God consolidated all legislative, judicial, and executive powers in the hands of a single deity (fig. 5). Though an individual might appeal to God—by praying to God or even by "arguing" with him as Abraham did to spare Sodom and Gomorrah from total destruction—there was no higher authority or principle that God himself was answerable to.

Figure 5: View of the mountain traditionally known as Mount Sinai, where the Bible says Moses received the Ten Commandments from God. From Henry C. Fish, Bible Lands Illustrated: A Pictorial Handbook of the Antiquities and Modern Life of All the Sacred Countries *(Hartford, CT: American Publishing Company, 1876).*

Furthermore, by identifying God as the creator of the universe and all living things, Hebrew monotheism made him the author of all natural phenomena. The making of nature, however, was not viewed as an end in itself but rather as the cosmic prelude to God's consummate act of creation, the making of man. As God's commandments would make clear, this final act of creation had a moral intent—to create from matter a being fashioned in God's own spiritual image, a being that would be fruitful and multiply and live in keeping with his divine commands.

God himself, however, was not bound by laws governing the cosmos. He could arbitrarily intervene in the otherwise natural order of things by manifesting what the Bible calls "signs and wonders," miraculous acts that defied the normal course of events. Thus, for example, he could part the sea,[5] make the shadow on a sundial reverse itself,[6] and even cause the sun to stand still in the sky.[7]

This authoritarian component of Hebrew theology tended to discourage and inhibit independent thinking. Thus Adam and Eve were forbidden to taste the fruit of the Tree of Knowledge[8] and, when they disobeyed God's edict, were severely punished[9] by being exiled from Eden, a paradise that could have only been enjoyed by those willing to stifle their curiosity.

Humanity's mission on the earth was not an intellectual but a moral one, the Bible said, for what God required was unquestioning obedience. The nearest thing to "research" was the diligent study and dutiful practice of God's ethical teachings. Truth was not something yet to be discovered but something that had already been revealed. As the book of Proverbs declares: "The beginning of wisdom is the fear of the Lord, and knowledge of holy things is understanding."[10]

For its part, nature, which celebrated God's majesty, did not merit investigation in its own right. Instead, the world of nature was merely the scenic backdrop for a divinely directed morality play acted out on the stage of history. Even a disease like leprosy was seen not as a malady crying out for scientific treatment and cure but as a curse visited upon humans by God.

The Bible's anti-intellectual bias is evident in the book of Job. A pious man, Job was tortured by God in order that God might win a bet with Satan. In spite of the loss of his fortune, the death of his children, and ravages of disease, Job refused to admit he was guilty of some sin for which he was being punished. Maintaining his innocence, he demanded to know why God had made him suffer. God's callous reply was that a mere human like Job had no right to demand an explanation, seeing that his powers were so pathetically feeble and his knowledge so utterly meager compared to those of the Lord. "Where were *you*," asked God haughtily, "when I laid the earth's foundation?"[11]

All Job could do in response was to humbly confess his ignorance.[12] "I know that You can do everything, that nothing is beyond your imagination.... Indeed, I spoke without understanding, of things beyond me which I did not know . . . I recant and relent, since I am but dust and ashes."[13]

The characters of the Old Testament, however, were not all cut from the same cloth. One character who came the closest to being an independent thinker was Koheleth, the author of the book commonly called Ecclesiastes.

To Koheleth's jaded eye, however, pursuing any goal in life was akin to chasing the wind. "Utter futility," he muttered over and over again. "Everything is futile!"[14]

Among such futile pursuits, he charged, was the pursuit of knowledge.

"The more wisdom, the more frustration. He who increases his knowledge only increases his heartache. . . . Making books is endless, and continual studying only wears out the flesh."[15] In the end, man cannot find the answers he seeks. "Man cannot discover everything that happens under the sun. For however much he may try, he will not find out."[16] Nor is the wise man superior for all of his striving. "The wise man will not be remembered forever, no more than is the fool; for, as the days roll by, everything will be forgotten. In the end, the wise man dies, just like the fool."[17]

Koheleth's cynicism complements Job's submissiveness, for both figures concluded that ultimate truths were beyond human reach. In effect, the angels that God had long ago stationed at Eden's border to block man's return were still there, wielding their flaming swords.

The religious values of ancient Israel thus tended to inhibit critical thinking and free inquiry, especially into the workings of the universe. Such knowledge, it was assumed, was not really important; in fact, it was beyond human comprehension. In addition, the Hebraic focus on the spiritual dimension of life served to de-emphasize the material world, which is the fundamental province of science.

For scientific progress to flower, a new people with new and different cultural values would have to emerge in the Mediterranean world, a people who would build on earlier discoveries and dare to reclaim Eden's forbidden tree. That unique people would be the Greeks.

Chapter 3

THE LANGUAGE OF THE UNIVERSE

"Philosophy is written in this grand book—I mean the universe—which stands continually open to our gaze, but it cannot be understood unless one first learns to comprehend the language and interpret the characters in which it is written. It is written in the language of mathematics, and its characters are triangles, circles, and other geometrical figures, without which it is humanly impossible to understand a single word of it; without these, one is wandering about in a dark labyrinth."
—Galileo Galilei, *The Assayer*, 6.
Translated by Stillman Drake and C. D. O'Malley.

According to a story, the Greek philosopher Aristippus was once shipwrecked with his companions on an unfamiliar and desolate shore.[1] As he wandered despondently on the beach, Aristippus suddenly came upon geometric diagrams drawn in the sand. "Take heart, my friends," the philosopher shouted, "for I have found the footprints of man!" (See fig. 6.)

The "footprints" that Aristippus had discovered were proof to him of humanity's presence, for geometry and mathematics are a uniquely human language. As the ancient Greeks recognized, they are also the language of the universe.

Greek science was based upon three fundamental assumptions: that behind all natural phenomena there exists order, that this order is intrinsic and not arbitrary, and that it can be discovered by the human mind.

The first assumption—that a hidden order underlies nature—is, in one respect, strikingly naive because it rejects the possibility that things may happen by sheer chance. Such a belief may have originated in prehistoric times, giving comfort to the earliest humans by banishing the fear of an inexplicable universe. Such a view probably also led to the birth of primitive gods who were said by priests to control these phenomena and who could be placated by ritual and prayer.

43

Arifippus Philofophus Socraticus, naufragio cum ejectus ad Rhodienfium litus animadvertiffet Geometrica fchemata defcripta, exclamaviffe ad comites ita dicitur, Bene fperemus, Hominum enim veftigia video.
Vitruv. Architect. lib.6. Praef.

Figure 6: An eighteenth-century engraving showing the philosopher Aristippus discovering the "footprints of man." From Thomas Heath, Archimedes *(London: Society for Promoting Christian Knowledge; New York: Macmillan, 1920).*

The second assumption—that such a natural order is innate—is more exceptional, challenging as it does the traditional notion of willful and at times capricious gods who rule the universe. In the ancient Mediterranean world, only the Egyptians had believed in a cosmic principle of justice and order (called *ma'at*) to which even the gods themselves had to conform. Before the dawn of science (and indeed for some time thereafter), the Greeks, like the Babylonians, worshipped gods whose personal behavior, as depicted in myth, was no different or better than that of mortals. Remarkably, Greek science would break with that mythic precedent and the religious traditions of the past.

The third and distinctively Hellenic assumption—that men can discern nature's hidden order—was the most revolutionary of all because it substituted the intellectual curiosity and autonomy of the individual for mindless obedience to a priestly elite. The means to discover nature's secrets became human reason, whose purest expression was the language of mathematics. Thus man's own rational nature came to articulate the rational nature of the cosmos.

The earliest civilizations of Egypt and Mesopotamia had excelled in mathematics and developed complex principles of arithmetic and geometry centuries before the Greeks. But the focus of these Near Eastern civilizations was on the practical applications of these principles for measurement in such fields as agriculture, commerce, and construction. Ancient Greek thinkers, however, were more fascinated by the abstract beauty and mystery of numbers and shapes rather than their practical application. The Greeks recognized that, were it not for numbers, nothing could be fully understood or described. And so, as Aristoxenus once said, "they took numbers out of the marketplace and honored them for themselves."[2] Among the earliest to do this was the sixth-century BCE mathematician Pythagoras, who was the first to call the universe a "cosmos," a word that implied both beauty and order and signified a harmonious arrangement of parts. His followers revered the "tetractus" (from the Greek word *tetra*, "four"), a mystical symbol consisting of the four alternating odd and even numbers 1, 2, 3, and 4, whose sum was the perfect number 10.

By a disciplined and logical progression that proceeded by deduction from basic definitions and self-evident truths to revolutionary theorems and corollaries, Greek mathematicians explored and mapped the primordial landscape of physical reality. The essential building blocks of the universe were their sole preoccupation—squares, circles, triangles, cubes, spheres, and cylinders—and the timeless commensurability of their shapes, surface areas, and volumes. Instead of equations, the common tools of modern mathematicians, ancient Greeks like Archimedes and Euclid traditionally used concise verbal statements illustrated by lettered diagrams, with algebraic formulas coming to the fore only in the third century CE. Because of their dedication to reason, the Greeks' greatest conceptual struggle was with the indefiniteness of irrational numbers and the idea of infinity, both of which seemed to defy the precision of logic and the notion of a finite universe that could be fully comprehended by man.

The basic questions that these Greek thinkers posed had but one aim: to satisfy their thirst for knowledge about the structure of the space man inhabited. Eventually, they applied their rigorous methods of investigation to study not only space but also movement through space, as they speculated on the passage of visual images and sounds, the motion of objects on Earth, and the transit of heavenly bodies in the skies above. In the end, they would also explore the universe within as they examined the nature of life and the inner mysteries of the human body.

It is now time for us to follow that great adventure and discover where those "footprints in the sand" would lead.

PART II:

░░░░░░░░░░░░░░░░░░

EXPLORING THE UNIVERSE

░░░░░░░░░░░░░░░░░░

SECTION 1: THE WORLD OUTSIDE

Chapter 4

OPTICS

Optics, the science of sight, came naturally to the knowledge-hungry Greeks. As their language proves, they singled out the sense of sight as learning's primary organ, for in ancient Greek "to know" (*idein*) meant "to have seen."

Yet before the ancient Greeks had scientists to study sight, they had wartime heroes to gauge its precious value.

The value the Greeks assigned to vision is evident in the two oldest literary works of Western civilization, the *Iliad* and the *Odyssey*. Ascribed to a blind Greek poet named Homer, these two epics date to the eighth century BCE but embody tales far older. The *Iliad* recounts Achilles's role in the Trojan War; the *Odyssey*, the postwar ordeal of a veteran named Odysseus who struggled to get back home.

Our first illuminating extract comes from the *Iliad* and ironically features a hero who was anything but a scientist or intellectual: a brawny Greek killing-machine named Ajax. During a fierce battle that raged between the Greeks and their Trojan adversaries, the battlefield suddenly became shrouded in darkness. Frustrated at his inability to kill the enemy, Ajax raised a prayer to Zeus, the king of the gods: "Father Zeus, deliver us from darkness and make the sky bright. Allow our eyes to see. If slay us you must, then slay us in the light."[1] (Note: All translations in this volume are the author's own, unless otherwise indicated.)

The words of Ajax illustrate an essential trait of Greek character: a compulsion to see and understand, even at the price of life itself. It was this abiding trait that, centuries later, made science possible and made the ancient Greeks its true inventors.

Our second illustration is from the *Odyssey* and stars the epic hero Odysseus, known in later times as Ulysses. Unlike his brawny comrade Ajax, Odysseus was a brainy warrior and a thinking man's hero, ever ready to use his cunning intelligence and protean imagination to defeat an enemy or escape from danger.

It was Odysseus, for example, who devised the stratagem of the Trojan Horse. When the city of Troy refused to fall to frontal assault, Odysseus determined to capture the city by deception. The scheme he concocted was to construct a giant horse out of wood, fill it with Greek commandos, and trick the naive Trojans into taking it inside their fortress. The unseen danger that lurked within the horse would prove to be Troy's undoing.

When the victorious Greeks sailed from Troy, Odysseus had many dangers yet to face. One of his most perilous adventures involved an encounter with a cannibalistic one-eyed giant named Polyphemus.[2] Odysseus and his men found themselves trapped inside the giant's cave facing almost certain death, and a grisly one at that. Only Odysseus's intellectual resourcefulness saved their lives. Getting the giant drunk, Odysseus introduced himself as "No Man." When the giant passed out, Odysseus and his men blinded him with a sharpened stake. As the Cyclops cried out in pain to his neighbors to come to his aid, he yelled, "No Man hurt me!"—and his fellow giants promptly returned to their respective caves convinced that no real harm had been done. The next morning, when the now-sightless Cyclops removed the boulder from the cave's mouth to let his flocks out to graze, Odysseus and his companions stealthily made their exit among the sheep. To add insult to injury, once onboard his ship, Odysseus triumphantly shouted out his true name so the vanquished monster would never forget it.

During his time trapped inside the dark cave, Odysseus had truly been No Man—vulnerable and impotent. But by drawing upon his inner resources of intellect and courage, Odysseus was able to reclaim his identity and stand proudly in the liberating light of day as "Someone" once again.

The admiring guardian angel who stood beside Odysseus through all his adventures was Athena, the Greek goddess of intelligence (fig. 7). Mythology tells us that her gift to humanity was the olive tree, a gift deemed precious not only because the olive became a staple of the Greek diet but also because it fueled the lamps of ancient Greece. The goddess of intelligence had thus given humanity a gift that could banish darkness, including the darkness of ignorance that could otherwise make humans powerless.

The poet Homer tells us that during his ten-year-long odyssey at sea, Odysseus "*saw* many cities of men and *learned their minds.*"[3] The most significant words here for the history of Greek science are the ones I have set in italics. Odysseus was no casual tourist, snapping pictures (had he had a camera) and picking up souvenirs. Instead, he was a true traveler, fascinated

Figure 7: Athena, the Greek goddess of intelligence, portrayed on a fifth-century BCE Athenian coin. From George Redford, A Manual of Sculpture *(London: Sampson Low, Marston, Searle, and Rivington, 1882).*

by the thinking and lifestyles of the cultures he engaged. He was a man whose brain soaked up experiences like a sponge, thirsty to understand the workings of the world he explored.

Though Odysseus's homecoming took a decade, the news of Troy's fall took only minutes to reach Greece, if the word of the fifth-century BCE dramatist Aeschylus is to be believed. The playwright tells us that Queen Clytemnestra of Mycenae set up an ingenious system of beacon fires across the 250 miles of land and sea that separated her palace in Greece from Troy's citadel.[4] Her purpose, however, was not dispassionately scientific. Instead, she wanted to know as soon as possible when her husband Agamemnon was returning from battle so she could set into motion her plot to murder him. Technology, we see, even ancient technology, can be used to achieve destructive ambitions.

The Nature of Vision

Between the days of Homer and Aeschylus, the ancient Greeks began investigating the nature of sight.

For most civilizations it would have been enough just to see, but the Greeks were not like most other civilizations. Where others simply took things for granted, the Greeks wondered and questioned as had no other culture. In the ancient Near East, in fact, religion had long been a barrier to the scientific investigation of natural phenomena. Whereas in ancient Egypt the sacred eye of Horus, worn as an amulet, was thought to magically protect the wearer, and in ancient Mesopotamia signs seen by a diviner were believed to prophetically disclose the future, in neither land was the actual process by which humans beings see subjected to rational scrutiny.

As the philosophers of Greece speculated on the nature of vision, they proposed two diametrically opposed models to explain its operation. According to one theory, objects transmitted images to the eye; according to the second, more popular theory, the eye itself generated rays that in turn struck objects.

In support of the first theory, Empedocles (ca. 492–432 BCE) argued that objects sent out emissions, or effluences, that were absorbed by pores inside the eye. On the other hand, Democritus (ca. 460–370 BCE) argued that these emissions were in fact tiny particles or, as he called them, "atoms." Later, Epicurus (ca. 340–269 BCE) claimed that the images—whatever they were—traveled at an incredible speed and moved as rapidly as thought.

Taking a totally different stance, Plato (ca. 429–347 BCE) in his philosophical dialogue the *Timaeus*, proposed that vision was a stream of pure, fiery energy that flowed from the eye until it met an object bathed in light, from which it then flowed backward to the eye that was its source, ultimately reaching the psyche, or soul. The later mathematicians Euclid (ca. 325–250 BCE) and Ptolemy (second century CE) supported Plato's underlying argument. In their view, the human eye—like radar—emitted rays that fanned out and traveled in straight lines until they encountered objects they recognized.

In the fifth century BCE, Alcmaeon of Croton had proposed that passages connected the sense organs, like the eyes, to the brain. Eventually, by performing autopsies on cadavers, Herophilus of Chalcedon (ca. 330–260 BCE) would discover just such a passage by dissecting the human eye, identifying the optic nerve, and tracing it to the brain.

Herophilus's work, of course, didn't settle the basic philosophical argument: there was still no way to prove which way the neurological traffic flowed—from the object to the eye or from the eye to the object—but at least science was moving beyond merely theorizing.

Before Herophilus's experiments, after all, the debate over the nature of vision had been just that, a debate. In the 400s BCE during Athens' "Golden

Age," gentlemanly Greek thinkers had been more inclined to rely on the pristine purity of logic than the nitty-gritty of actual experimentation.

Even so, the value of such early philosophical debate must not be underestimated, for impassioned disagreement, however abstract, is part and parcel of what science, including modern science, is all about.

Athena's gift was wisdom, but sometimes wisdom doesn't come easy. Athena's first worshippers, the ancient Greeks, had conceived the radical notion that reason, not divine revelation, was the surest pathway to truth, yet they simultaneously acknowledged that the road was an inevitably bumpy one. Lovers of both liberty and logic, they reveled in free debate, believing that only through the dialectic give-and-take of argument could truth be most likely discovered, even as a blade can only be sharpened if it is honed against an opposing surface.

In fact, the ancient debate over the mechanisms of vision would anticipate similar scientific debates over two thousand years later concerning the nature of light. Thus, in the eighteenth century, most scientists were convinced that light was composed of particles; by the nineteenth century, they realized that light might be composed of waves; and by the twentieth century, they concluded that light traveled in waves that were in fact made up of particles. Yet without the Greek penchant for examining first principles and courageously debating their merits, such scientific progress would have been impossible.

The Greeks' exploration of vision focused on four phenomena: reflection, refraction, perspective, and optical illusion.

A QUESTION OF REFLECTION

Besides being curious about vision in general, Greek thinkers were fascinated in particular by the phenomenon of reflection. This phenomenon, in fact, inspired one of mythology's most poignant and enduring tales, the Greek legend of Narcissus.

According to the story, Narcissus was an extraordinarily handsome young man. Though desired by many lovers both male and female, he callously spurned them all. In frustration, one of them cursed Narcissus, praying that he should someday suffer the same pangs of rejection his lovers had known.

Later, as Narcissus was walking through a wood, he chanced upon a pool and bent down to refresh himself in its waters. As he leaned over, Narcissus saw a lovely face looking up at him. Not realizing it was his own reflection, he spoke to it and drew closer to the water's surface, begging the image to come out, but to no avail. Its perpetual refusal only served to intensify his erotic desire. Continually tormented, Narcissus remained by the edge of the pool, not eating or drinking, until he wasted away and perished from unrequited love. From this tragic story the word *narcissism* arose to describe the tragic self-love from which he died.

The effects of reflection were not merely the subject of myth but also became a topic of serious speculation among Greek scientists.

Around 300 BCE, Euclid demonstrated by incontrovertible logic that the shortest path between two points must be a straight line. In the first century CE, basing his calculations on Euclid's studies of triangles, Heron went on to prove that the angle of incidence on a flat mirror equaled the angle of reflection, that is, the angle formed between a mirror and an observer's line of sight equaled the angle from the mirror's surface to the object observed. (Ancient Greek mirrors, we should note in passing, were made not of silver-backed glass as they are today but of polished copper or bronze.) A century later, Heron's theoretical model of reflectivity was confirmed in laboratory experiments by his fellow Alexandrian, Ptolemy.

Heron's work is the oldest complete treatise on mirrors to survive. In the following passage, he catalogs a variety of unusual applications for a mirror's reflective surface:

> Reflectivity not only merits scientific investigation but generates effects that can dazzle the eye. Scientifically constructed mirrors can make the right side of an object appear on the right, and the left side on the left, whereas ordinary mirrors reverse an object's sides. Mirrors can also let us see our own backs, make us look like we're standing on our heads, give us three eyes or a couple of noses, or simulate sorrow by contorting our faces. But the science of reflection, apart from the spectacular effects it can produce, can also have important practical applications. While staying inside our own houses, for example, we can secretly see who's outside and watch what they're doing. In addition, we can tell what time it is day or night by using a clock fitted with mirrors that can show us different images that rotate as the hours go by. There are even mirrors that don't simply reflect the person standing in front of them but can be made to project whatever image their operator wants.[5]

Greek scientists in Alexandria, Egypt, would go on to devise complex mathematical formulas to describe how images were formed on mirrors with curved surfaces. As early as 200 BCE, a Greek mathematician named Diocles had written a treatise titled "On Burning-Mirrors." Commenting on conic projections, Diocles had described the hypothetical construction of a parabolic mirror that could focus its rays and set targets ablaze.

During the Hellenistic era, Alexandria was the cultural capital of the Mediterranean world. Alongside the tomb of Alexander the Great, the monumental "crown jewels" of the city were its renowned research facility for scientists and scholars, known as the Library and Museum, and the city's magnificent Lighthouse (fig. 8). Named for the rocky island upon which it stood, the "Pharos" guided ships to Alexandria's spacious and welcoming harbor. Erected at the beginning of the third century BCE under Ptolemy I and II (no relation to the mathematician of the same name), the lighthouse came to be celebrated as one of the Seven Wonders of the World. Of all these monumental wonders—the Great Pyramid, the Hanging Gardens of Babylon, the Statue of Zeus at Olympia, the Mausoleum of Halicarnassus, the Temple of Artemis at Ephesus, and the Colossus of Rhodes—the Pharos was the newest and, given its nautical mission, the most practical.

Figure 8: Reconstruction of the Lighthouse of Alexandria. From Hermann Thiersch, Pharos: Antike Islam und Occident: Ein Beitrag zur Architekturgeschichte *(Leipzig and Berlin: Teubner, 1909).*

Because Egypt's coastline is so flat, Mediterranean mariners who wanted to avoid its dangerous reefs and shoals needed a tall landmark to steer toward. The structure designed to meet their needs rose to an estimated height of over three hundred feet and was constructed of white stone (limestone, marble, and/or granite) that would shine brightly in the Egyptian sun. Rising from a 200-foot-tall square base, a 100-foot hexagonal tower climbed to support a third cylindrical story about 50 feet high, which was in turn surmounted by a statue of the god Poseidon, savior of sailors, possibly gilded to better reflect sunlight. On the four corners of its base, sculpted Tritons—mythical sentinels of the sea—symbolically blew on conch shells, the "foghorns" of ancient times. Well over half the height of the Washington Monument, the Lighthouse of Alexandria was in its era the tallest structure ever built in the ancient world, a skyscraper second in height only to two of the three major pyramids at Giza. Its design may have been inspired by the obelisks of Egypt, some a hundred feet tall, that were plated in gold to reflect the rays of the sun god Ra whom they honored.

Within the lighthouse, a broad flight of stairs on the lowest level led to a narrower staircase that spiraled to the top, with steps illuminated by multiple windows piercing the walls. Mounted in the topmost section of the building (possibly as a dramatic afterthought) was a furnace. Backed by a curved and polished bronze mirror, its fire blazed so fiercely at night that its glow could be seen by ships thirty miles out to sea, while during daytime the mirror shimmered in the light of the sun. "By night it resembled a gleaming star," wrote one visitor of the Pharos, "and by day we could see its rising smoke."[6]

How the fire was fed is itself a marvel, since supplies of timber in Egypt have always been meager. The fuel of choice may have been imported wood or olive oil, or even dried manure. Yet the manner in which the fuel was lifted to the top of the 350-foot-tall tower is just as much a mystery. Of course, the furnace could have been located at a lower level, but then the light from its fire would have to have been somehow transmitted to the summit, perhaps by a series of mirrors, albeit mounted inside a chimney filled with smoke.

The designer of the Pharos was named Sostratus: either a Hellenistic engineer or a wealthy Alexandrian courtier of the same name. According to what may be an apocryphal tale, the real Sostratus engraved his name on the monument, covered the writing in plaster, and then inscribed his king's name over it, knowing that with the passing of time Ptolemy's name would vanish, and his own would remain forever, eternally declaring his glory to all the world.

As a Wonder of the World, the Pharos was pictured on mosaics and coins and even on sarcophagi, becoming a model for modern lighthouses and for other similar ones built around the ancient world, including one at Ostia, the seaport of ancient Rome (fig. 9). After standing proudly for fifteen centuries as a beacon to mariners and a tribute to architectural and engineering genius, the Pharos collapsed in the 1300s, the victim of devastating earthquakes.

Figure 9: Roman Imperial coins depicting the then-standing Lighthouse of Alexandria. From Hermann Thiersch, Pharos: Antike Islam und Occident: Ein Beitrag zur Architek-*turgeschichte (Leipzig and Berlin: Teubner, 1909).*

Besides helping sailors in distress, a mirror could also serve as a deadly weapon, Greek mythology reveals.

In mythology, Medusa was a Gorgon, a horrific creature whose look could turn a man to stone. Vowing to slay the beast, the hero Perseus received a special gift from Athena, a highly polished bronze shield, and special instructions in how to skillfully employ it. Having discovered Medusa's lair, Perseus silently crept up behind her, averting his face to avoid looking into the monster's eyes and gazing instead at the reflection of her head as he stealthily grew closer. Finally, when he was near enough, he swung down his sword and decapitated the Gorgon.

The value of a mirror in combat, however, was not limited to myth, for there was a time in ancient history when—thanks to scientific ingenuity— mirrors were put to use in war.

During the third century BCE, the Romans vied with the North African city of Carthage for control of the western Mediterranean. Though previously defeated by the Romans in Africa, the Carthaginians decided to engage the Romans once again by crossing the Alps under Hannibal and invading Italy from the north. The rich and powerful Sicilian city-state of Syracuse, previously allied with Rome, now decided to throw its support to Carthage. To force the Syracusans back into submission, the Romans in 213 BCE sent a combined land and naval force to besiege the city.

Syracuse had been founded centuries before as a Greek colony and had been ruled by monarchs who traced their lineage back to Greece. Following the first Punic War that saw the defeat of Carthage, the third-century BCE Greek mathematician Archimedes settled in Syracuse and worked under the patronage of the city's king. Anticipating a Roman invasion, he lent his talents to helping his adopted homeland prepare for attack.

Perhaps the most spectacular weapon Archimedes is said to have invented was a "death ray," a laser-like device to burn the Roman battleships that sailed into Syracuse's harbor. The tale was told by the Greek historian Cassius Dio (ca. 164 to ca. 229 CE) in his comprehensive account of Rome's rise to power. In Book 15 of his *History of Rome*, he described the siege of Syracuse and Archimedes's role in repelling the Romans. Regrettably, that chapter of his work has been lost, but fortunately, two twelfth-century Byzantine summaries of it survive. One version says: "In the end, Archimedes succeeded in burning up the entire Roman fleet in an amazing way. Tilting a type of mirror in the direction of the sun, he concentrated the sun's rays on it. Due to its thickness and smoothness, the mirror ignited the very air, producing a great flame that he then directed at the ships lying at anchor until they were all consumed by fire."[7]

According to a second, more detailed account: "When Marcellus [the Roman admiral] drew his ships out of the range of the defenders' arrows, old Archimedes set up a sort of hexagonal mirror, arranging smaller squarish mirrors at intervals that were hinged and moveable by a chain. He then focused the rays of the noonday sun on the device and ignited the ships with awesome effect, turning them into ashes at the distance of a bow-shot. Thus did the old man use his weaponry to defeat Marcellus."[8]

Though these Byzantine summaries agree on the main points of the story, they date to the twelfth century CE, over a thousand years after the days of Cassius Dio, the writer whom they quote and who was himself not an eyewitness to the events he described four centuries after they took place. Apart from any suspicion we might have about the reliability of such late sources, the notion of an ancient solar-powered "death ray" would strain our credulity—except for the fact that Archimedes's contemporary, Diocles, had already theoretically described just such a device.

In 1973, with abiding faith in the veracity of the Archimedes legend, a modern Greek engineer named Ioannis Sakkas actually succeeded in duplicating his ancient mentor's feat. Assisted by seventy sailors from the Royal Greek Navy who held an array of mirror-like metal shields, Sakkas was able to set fire to a wooden ship. His stunning achievement was documented on film.

In 2005, the same feat was repeated by a group of student volunteers from MIT led by their professor, David Wallace. The MIT team bought 127 inexpensive one-foot-square silver-backed glass mirrors and arranged them in a two-tier arc on the roof of a campus building. One hundred feet away they set up a wooden mockup of the side of a ship. In ten minutes the students achieved a flash ignition, and the ship's hull burst into flames.

Later invited by the Discovery Channel's *Mythbusters* show to duplicate their feat under real-world conditions, the MIT team set up four tiers of one-foot-square tiles made of polished bronze and aimed them at a wooden boat floating 110 feet away. The concentrated solar rays were able to char the wood and burn a ten-inch hole in the hull, but they did not set the boat afire, perhaps—the team reasoned—because their target was a refloated derelict fishing boat whose wood was still damp.

As Professor Wallace readily admitted, the experiment didn't prove that Archimedes had ever actually constructed or used a solar "death ray," but it did demonstrate that such a device could have worked, at least under controlled conditions. Whether it could have effectively focused its deadly rays on ships that were in motion—bobbing up and down at anchor or being rowed—is even more difficult to say. Syracuse's harbor faced the east, and Archimedes would have needed a clear sky as he aimed the device into the late morning sun. At the very least, the brilliant glare from his mirrors would have blinded the Roman commander and his crew as they readied their attack.

Despite Archimedes's valiant efforts, the Romans eventually succeeded in breaching Syracuse's walls. During the subsequent looting of the city, Archimedes himself died tragically, killed when he ignored a Roman soldier's command to stand up. He was preoccupied, the story goes, drawing a geometric diagram in the sand.

BENDING LIGHT

Our best documentary source on the bending of light, or refraction, is Ptolemy's book *Optics*, which dates to the second century CE. Ptolemy not only discussed the mathematics of refraction, but he also confirmed his theories through tabletop experiments, carefully testing what happens when light passes from one medium to another of differing density: from air to water, from air to glass, and from water to glass. He found that as light travels into a medium of greater density, it ceases to move in a straight line but instead bends, angling away from a plane perpendicular to the medium's surface. Using a special disc for sighting that he calibrated in degrees, he then measured how much an underwater object seemed to bend as his line of sight changed. (Recall that, for Ptolemy as for Euclid, "sight rays" supposedly emanated from the eye and traveled to the object.)

What Ptolemy was doing was systematically investigating a phenomenon known to the Greeks for centuries—and one that we can observe for ourselves. If a coin is placed at the bottom of a bowl near the bowl's edge, it seems to move closer to the center of the bowl when the bowl is filled with water. (To do the experiment you'll either need a heavy coin like the ones the Greeks used or you'll need to press the coin firmly to the bottom of the bowl with the eraser-end of a pencil so it won't actually move as water is poured in.) Ptolemy's contribution was to carefully calculate the shifting of the image so that for every angle of incidence, he could predict a corresponding angle of refraction and thus develop a universal formula that could apply to all situations.

That the Greeks applied the principles of refraction to their everyday life is illustrated by a surviving passage from an ancient comedy.

The master of classical Greek comedy was the fifth-century BCE Athenian playwright Aristophanes. In his play *The Clouds*, he ridiculed the philosopher Socrates and the way sophistic reasoning could undermine morality. As a case in point, Aristophanes introduced a slippery character named Strepsiades, who was trying to sidestep a lawsuit:

> STREPSIADES: I just found a really smart way to avoid getting sued. I bet you'll agree.
> SOCRATES: How?
> STREPSIADES: Well, have you ever seen that stone they sell in drug-stores—the pretty one you can see through, the kind they use to light fires?

SOCRATES: You mean the glass one?

STREPSIADES: Right! How about if, when the clerk is writing up the case on his wax tablet, I stand between him and the sun and melt the words right off the page?

SOCRATES: Couldn't have come up with a better idea![9]

This passage reveals that, two centuries before Archimedes's death ray, the Athenians were using magnifying glasses to set fires with the help of the sun. It's only natural to assume that they also would have understood the capacity of such a lens to magnify objects.

In each case light is refracted, or bent: When it is reflected from a surface, the light rays expand and enlarge the image that comes to our eye. When the light comes from a radiant source like the sun, the rays passing through the lens converge to a focal point and there concentrate their heat. But did the ancient Greeks understand this optical principle?

As far as we know, however, the ancient Greeks did not investigate the type of refraction that would have taken place in a curved lens such as Aristophanes's "burning-glass." Whether they ever deliberately produced lenses for the purpose of correcting or, more likely, enhancing vision is a matter of scholarly debate. Transparent lenticular objects survive archaeologically—some made of glass, others made of crystal, and still others of amber—but most were probably used as ornaments for artwork or jewelry rather than as aids for engravers seeking to achieve fine detail on a miniature scale.

GAINING PERSPECTIVE

Besides laughing at the comedies of Aristophanes, theatergoers in Athens were moved by thought-provoking tragedies written by such fifth-century BCE playwrights as Aeschylus and Sophocles.

Visually, ancient Greek drama was very conservative: actors wore traditional masks, and the stage was backed by the standardized facade of a mock palace or temple. But during the fifth century BCE, a painter named Agatharchus of Samos introduced a revolution in scenic design: a more naturalistic flat backdrop that was painted with an imaginary scene shown in perspective. While some parts of it seemed to stand out in the foreground, others looked like they were receding into the distance. Agatharchus was hired by Aeschylus and Sophocles and even went on to write a book, now lost, about his technique that came to be called *skenographia*, or scene painting.

Up until that time, the use of any perspective in Greek art had been very rare, apart from an occasional vase painting. Yet thanks to Agatharchus's innovation, more and more Greek artists began to experiment with its use. Sculptors, for example, who decorated temples with marble friezes carved in relief, used a technique called foreshortening, progressively reducing the dimensions of a human figure as it merged into the background to create an impression of greater depth. Greek wall painters may have even employed a single unifying vanishing point for all horizontal lines. Though many scholars attribute this achievement to the artists of the Italian Renaissance, Renaissance painters may have based their approach on earlier Greek insights into human vision, insights recorded in Euclid's enduring book on optics and preserved and transmitted to posterity by Romans such as Vitruvius. For example, Euclid's fifth proposition states: "Equal magnitudes situated at different distances from the eye appear unequal, and the nearer always appears larger," while the sixth declares: "Parallel lines when seen from a distance appear to be an unequal distance apart."[10]

OPTICAL ILLUSION

Euclid's observations imply that the eye can be fooled by what it sees, that there is a difference between what "seems" and what actually "is." In fact, in his later study of optics, Heron diagrammed two experiments involving the use of mirrors to create optical illusions based on this principle: one that would make an observer think he was flying, and another that would substitute the image of another person for the observer's own. But ancient Greece's most impressive optical illusion was a stunning work of architecture.

High atop the Acropolis overlooking Athens' great theater was the Parthenon, the largest temple ever built to honor Athena, the Greek goddess of reason (fig. 10). Constructed between 447 and 438 BCE after the devastating Persian Wars, the Parthenon celebrated Athens' victory over the foreign invader and the role Athena had played in defending her people and their way of life.

Athens' leading statesman, Pericles, chose two architects for the job, Ictinus and Callicrates. Sharing their city's passion for perfection, Ictinus and Callicrates wanted to make the Parthenon the most perfect temple ever built. But a major obstacle stood in their way: the nature of sight.

The eye itself contains a lens that bends, or refracts, light. As a result, when we look at an object in space, lines that are perfectly vertical or horizontal can seem to be bent out of shape. We call this phenomenon an optical illusion. In effect, because of the innate limitations of human vision, what is perfect in terms of physical reality may actually look imperfect to our eyes.

In some cases the illusion is caused by the observer's point of view, the most common example of this being perspective. When parallel lines retreat into the distance, they really *don't* merge at all but in fact stay parallel. The

Figure 10: A nineteenth-century view of the Parthenon. From Samuel G. Green, Pictures from Bible Lands with Pen and Pencil *(London: Religious Tract Society, 1879).*

impression we have of their merging is an illusion. In the same way, an object viewed from a distance may *seem* smaller than a similarly sized object up close, but in fact both objects are alike, however much they may appear different to our eyes.

Because of such distortion, if the line of the Parthenon's superstructure had been perfectly horizontal, it would have looked to the viewer below as though it were curving downward, or sagging, in the middle. Likewise, if the tall columns supporting the superstructure had been perfectly vertical, to the observer looking up from below, the tops of the columns would have appeared to be leaning toward him, suggesting they might topple and cause the whole structure to collapse. In addition, if the sides of each column had been perfectly parallel, they would have seemed to bend in toward each other, making the column appear weak in the middle and too frail to support its load.

To compensate for optical distortion and to counter the impression that the temple was unstable and its columns weak, the architects deliberately bowed the superstructure up in the middle so that it would in fact look perfectly horizontal to the eye. They also tilted the columns inward toward the temple's interior to make them appear to stand straight and—following a traditional architectural practice already a century old—made sure the sides of the columns bulged outward to strengthen the impression that they were solid.

Indeed, it has been said with accuracy that there is not a single straight line in the entire Parthenon. Everything is curved. In fact, the whole floor of the building swells like a giant convex lens.

With these and other subtle refinements, undetectable to the untrained eye, Ictinus and Callicrates achieved their intellectual goal of creating an image of perfection. In actuality, they were not building a temple on the Acropolis; they were constructing it holographically in the observer's mind. By the mid-fifth century BCE, classical architecture had become a mind game in which creating and projecting a satisfying illusion was deemed superior to accepting the limitations of reality.

In fact, all of Athens was playing the same mind game, mistaking illusion for truth in their civic lives. During the Golden Age, the Athenians had convinced themselves they were godlike and invincible and could escape any consequences for their flaws. Armed with such an ideology, inflated by affluence, and intoxicated by the ideal sculptural images of human perfection they saw around them, they became an imperialistic state, despotically dictating their wants to the rest of Greece until their subject states rebelled under the leadership of Sparta. The long war that followed, from 431 to 404 BCE, saw the Athenians fall from the pedestal of moral leadership they had once occupied. This time, not even Athena could save them.

Some playwrights and philosophers had understood the dangers of such civic self-delusion. They had warned their fellow citizens of the risks of self-deception and the consequences of believing that the ends justify the means.

Looking back and remembering the execution of his beloved mentor

Socrates, Plato cautioned his fellow Athenians about the power of prejudice and the need to courageously look for the truth. To illustrate this in *The Republic*, he told a parable about shackled prisoners whose only reality were deceptive shadows projected by their masters onto a wall. To see the world for what it truly was, Plato tells us, the prisoners had to break free of the chains of blind ignorance and struggle toward the light. Other philosophers went so far as to criticize the theatrical art of *skenographia* as a deliberate effort to distract the public from a deeper reality they ought to face. As Sophocles showed in his tragedy *Oedipus the King*, the blind prophet Tiresias could see far more than the sighted King Oedipus, for our eyes and even our minds can delude us.

OPTICAL EFFECTS

There are a variety of ways we can gauge the influence of an ancient science like optics upon our times. First, we can single out modern technologies that have been directly inspired by ancient optical inventions. Second, we can identify technologies that were absent in antiquity but owe their very existence today to the speculations and experiments of the ancients. And last, and perhaps most important, we can point to modern ways of thinking that have been influenced, or may yet be influenced, by the scientific pathway of the ancient Greek mind.

To the first category—the legacy of technology—belongs the high-flux solar furnace, the origin of which can be traced to Archimedes's legendary "death ray." Located on a desolate mesa overlooking Golden, Colorado, the first such furnace was constructed two decades ago by the National Renewable Energy Laboratory (NREL). At its heart was a large motorized mirror, with a flat surface thirty-two meters square, that tracked the path of the sun as it crossed the sky and reflected its bright image onto an array of twenty-five smaller mirrors that were curved. These curved mirrors in turn concentrated the sun's rays into a beam that could generate a heat equal to fifty thousand suns. When employed commercially, such environmentally clean furnaces can have major industrial applications, including the flash coating of metals and ceramics, the accelerated testing of weathering, and the rapid detoxification of hazardous wastes. The basic design of Archimedes's destructive death ray can thus be turned to constructive use.

With far wider functions are modern technological applications of ancient experimental discoveries. Though there were no optometrists or opticians in ancient Athens, it was the Greeks' scientific curiosity about the refraction of light that later helped make the correction of vision through eyeglasses and contact lenses an everyday reality. Their discoveries, moreover, led to the development of lenses, which then led to the microscope and telescope, thereby exponentially increasing the power of human beings to explore the universe. An understanding of differential refraction has also made fiber optics feasible, a technology based upon the modern discovery that impulses of light can be guided along transparent filaments as long as they are surrounded by other materials with a lower refractive index. Thus fiber optic messages can be transmitted today along flexible cables at nearly the speed of light, much as thousands of years ago in a more primitive way a chain of signal fires would let Clytemnestra know her royal husband was returning from Troy.

The third category—the ways the ideas of the Greeks influence our thinking—can best be illustrated by their understanding of optical illusion. In their scientific search for truth, the best thinkers of Greece appreciated the critical importance of sight but distrusted the axiom that "seeing is believing," for they knew how readily the mind can be fooled by what the eye is shown. In a world like ours, where so much of what we take to be true comes to us in the form of images on a screen, we would be well advised to beware the dangers of delusion, of basing our decisions on persuasive pictures artfully manipulated by others for their own selfish ends. After all, when the citizens of Athens came down from the Acropolis or exited the theater at its base, they physically left behind the seductive artistry of architectural and theatrical distortion. Our own culture, however, is electronically saturated with artificial images that totally suffuse and dominate the environment in which we live, images that therefore are all the more capable of deceiving us.

Chapter 5

ACOUSTICS

J ust as the word *optics* comes from the Greek word for "eye" (*ops*), so the word *acoustics* comes from the Greek word for "hear" (*akouo*). In each case, we are indebted to the ancient Greeks not only for the words they gave us but also for the fundamentals of the sciences derived from them.

THE SOUND OF MUSIC

That the Greeks should have had an affinity for sound is natural, for their language was intrinsically musical. Unlike most languages of the ancient world, where words were accented by pronouncing one syllable louder than others (as is still true in English today), accents in ancient Greek were expressed by variations in tone, whether rising, falling, or rising and then falling. Besides such differences, the syllables of Greek words were also distinguished by their length, that is, by how much time it took to pronounce them—a long syllable taking about twice as much time to say as a short one.

Thus, the ancient Greek language contained the basic components of music: rhythm and pitch. By arranging words in artful ways, the poets of Greece created rhythmical and tonal patterns, and even internal melodies, often singing the poems they had composed to the accompaniment of a lyre or flute. Though other ancient Mediterranean civilizations appreciated music and played instruments, only the language of the Greeks was so distinctive in its inherent musicality.

The Greek love of beautiful sounds is also evident in the role music played in their literature and mythology (fig. 11). In Homer's *Iliad*, the feared warrior Achilles strummed a silver lyre in his leisure and sang of the brave deeds of heroes, while in Homer's *Odyssey*, Odysseus was enchanted by the seductive voices of the goddesses Circe and Calypso and hypnotized by the songs of the deadly Sirens. When a nymph named Syrinx fled from the amorous advances of

Figure 11: Musician playing a lyre on an Athenian vase of the sixth century BCE. From A. S. Murray, Handbook of Greek Archaeology *(New York: Scribner's, 1892).*

the god Pan and tried to disappear by transforming herself into a patch of thick reeds, Pan cut the reeds down in unequal lengths and, binding them together, made a set of pipes for himself. In a similar way, we're told, Pan's divine father, Hermes, invented the lyre by making a sounding box, or resonator, out of a tortoise's shell and stringing it with the dried sinews of a cow. Yet of all the tales in Greek myth, perhaps none is more poignant than that of a nymph named Echo, who was punished for her talkativeness. Forbidden from uttering her own thoughts, she was forced instead to "echo" the last words spoken by others. To her regret, Echo later fell in love with and was spurned by the self-absorbed Narcissus.

But the ancient Greeks did more than just appreciate sound. Forever curious, they questioned its very nature. "What *are* sounds?" they asked. "How do they originate and how do they travel?" "Why," they wondered, "are some sounds loud and others soft, some notes high and others low?"

If we are to believe their stories—and who can be sure, given the scope of the Greeks' imagination and their insatiable need to explain everything—it all began one day in the fifth century BCE when philosopher and mathematician Pythagoras went for a walk. As he strolled through the quarter of his city where the tradesmen plied their crafts, he heard a loud clanging. Approaching a smithy, his ears picked up a variety of sounds as the workmen's hammers struck various pieces of metal. Some of the metallic "notes" were higher and others lower. "What would account for the difference?" he wondered—a question that non-Greeks had never bothered to ask.

Assuming at first that the difference in tone might be due to the fact that some metalsmiths were more muscular than others or wielded different-sized hammers, he politely asked them to exchange their work stations and tools. Though the workers obliged, perhaps curious themselves about the odd fellow's motives, in the end it made no difference. It was the sizes of the metal plates when struck that accounted for the variation in tone.

Back home, Pythagoras experimented with different types of materials and objects to see if he could produce the same effect. Remembering Pan's pipes and how each tube gave out a different note based on its length, he realized that the same was also true of lyres, where different lengths of string tended to produce different sounds. Using a single-stringed instrument, or monochord, Pythagoras experimented with sounds by moving the bridge and comparing the length of the string with the note it produced. By reducing its length by half, for example, he raised the tone it generated by a whole octave; by doubling its length, he lowered it an octave; and, when sounded simultaneously, tones separated by an octave generated a single harmonious sound. He even tried filling pottery jars with different amounts of water, so the story goes, to see if the different water levels would change the sounds that emanated from the vases when he tapped them.

Drawing upon these analogies, Pythagoras fashioned a general principle to explain all the cases he had observed. (We call such a process of moving from particulars to a general conclusion "inductive reasoning.") Pythagoras discovered that all the materials and objects he had studied were answerable to the same law, a universal law of sound that could be expressed through the language of numbers. Not only was there a regular and predictable progression of tones intrinsic to the design of every instrument, but the notes that made the most pleasant and harmonious sounds when played together (the octave, the fourth, and the fifth, as musicologists call them) were related by fixed ratios based on what we now know to be their vibrational frequencies. Meanwhile, a contemporary of Pythagoras, Hippasus of Metapontum, conducted similar experiments by striking a series of suspended bronze disks of different thicknesses.

Not only had Pythagoras discovered a universal law of nature; he had demonstrated two other momentous and, at the time, astonishing truths: nature behaves in an orderly manner, and its ways can be rationally discerned by the human mind and described in mathematical terms.

In the following century, a Greek scientist from southern Italy named Archytas applied simple common sense and reasoned that louder sounds must be sounds that are generated with more force, and that because of that, greater force can travel farther than softer sounds. He also speculated that the "speed" (we would call it the "frequency") of a sound determined its pitch. Other Greek thinkers would argue that air must be the medium of transmission between the point of a sound's origin and the ear that hears it. Drawing an analogy between air and water, still others would propose that sound travels through the air in waves that move outward from the point of origin until they strike the ear, not unlike the ripples that spread outward when a stone is dropped into a pond. And just as waves or ripples in a pond may rebound when they strike the shore, so can sound be reflected backward as an echo.

By using their power of reasoning and drawing upon analogies between separate events, Greek scientists stripped away the older, *mythical* explanations for phenomena (e.g., "When you hear an echo, you're hearing a nymph who's repeating other people's words") and replaced them with newer *natural* explanations. In short, the mythic imagination was being displaced by the imaginative power of science.

THE THEATER OF SCIENCE

The theaters of Broadway and Hollywood hark back to buildings first erected by the Greeks over twenty-five centuries ago. In seeking inspiration for their tragedies, the playwrights of ancient Athens most often turned to the annals of mythology as they explored the interaction between the gods and humans and the tension between destiny and freedom. But in designing the theaters themselves, the architects of ancient Greece turned for guidance to the science of acoustics.

The earliest theaters of Greece took advantage of natural settings: the curve of a sloped hillside for audiences to sit on and a flat piece of ground below where actors could stand and speak their lines and—since Greek theater was always *musical* theater—where a chorus could sing and dance. To provide for lighting, Greek theaters were set outdoors and open to the sky. As time went on, wooden bleachers were constructed, as well as a wooden building behind the cast to serve as an elementary stage setting and as a place where the actors could store their props, change into their traditional costumes and masks, and make their entrances and exits through doors. After the fifth century BCE, a raised stage was introduced.

From an acoustical standpoint, the resonant wood of the scenic backdrop made the voices of the actors and chorus reverberate and bounce off toward the audience. The masks, for their part, were formed with large open mouths internally shaped to function like megaphones to project the actors' voices.

Even more disadvantaged than Greek actors were Greek orators, who had to address huge throngs of citizens in open-air assemblies using the unaided power of the human voice, guided only by their skill in enunciation. Such speeches were delivered as far back as the days of Homer's heroes,[1] a practice made regular by the demands of Athenian democracy. According to stories told by Plutarch in his "Life of Demosthenes," the fourth-century BCE Athenian orator Demosthenes was coached in enunciation by an actor named Satyrus and he practiced aloud for months at a time in a resonant underground chamber, or "sound studio," he had built himself where no one could overhear him. He overcame his tendency to stammer, Plutarch reports, by reciting speeches with his mouth full of pebbles and developed breath control by delivering speeches while running or climbing steep hills. Furthermore, to build up the power of his voice so it could be heard above the roar of the crowd, he rehearsed speeches on the seashore against the sound of the pounding surf.

The Roman engineer and architect Vitruvius informs us that, in order to enhance acoustics, some Greek theaters were equipped with custom-made bronze vases, adjusted for pitch and positioned in the seating area to focus and amplify the sounds emanating from the outdoor stage. Though no such vases have ever been found, comparable earthenware ones have been discovered in their original locations.

As communities grew in wealth, public-spirited citizens helped finance renovations that transformed theaters made of wood into theaters built of limestone and marble—as always, still open to the sky. In fact, some of these ancient theaters had a capacity for as many as ten thousand to fifteen thousand spectators.

THE MYSTERY OF EPIDAURUS

The most famous and revered theater in ancient Greece was the Theater of Dionysus in Athens, dedicated to the patron god of drama and located on the south slope of the Acropolis. But the best-preserved theater in Greece is found southwest of Athens at Epidaurus (plate 2), a site dedicated to Asclepius, the god of healing, and the center of a religious complex not unlike Lourdes, dedicated to faith healing on a massive scale. Because Epidaurus's theater could hold over fourteen thousand visitors, perfect acoustics were an absolute must. Thanks to the skill of the fourth-century BCE architect, Polycleitus the Younger, such perfection was achieved, so much so that even today someone sitting in the topmost row—the fifty-fifth—can hear the faintest tapping of a coin or the faintest whisper coming from the center of the performance area below, even though the source of the sound is some 170 feet away.

For a long time it was thought that the perfectly symmetrical design of the semicircular seating area—shaped like the truncated cone of a gigantic loudspeaker—was the chief reason for these consummate acoustics (fig. 12). Others hypothesized that the prevailing wind in this open-air setting, blowing toward the audience from the stage, might have carried with it the sound of the actors' voices (though experiments have now shown that such a rushing of air would more likely muffle the sound). Only recently, however, with the aid of electronic detection equipment and data processing was the complete answer found by two Georgia Tech engineers, Nico Declercq and Cindy Dekeyser.[2]

They discovered that the auditorium's limestone seats, which were corrugated, or ridged, acted like a filter for sound waves operating at different frequencies. Bass frequencies below 500 hertz (such as the noise from the rustling of the audience in their seats) were suppressed or muted, while treble frequencies higher than 500 hertz (such as the voices of the actors or chorus) were allowed to register on the audience's ears. Compensating for the "missing" low frequencies that would have come from the stage was a neural mechanism known as "virtual pitch," by which the human brain automatically reconstructs low frequencies it would have expected to hear under normal circumstances, an effect that takes place today when we listen to the small speakers built into a laptop or telephone.

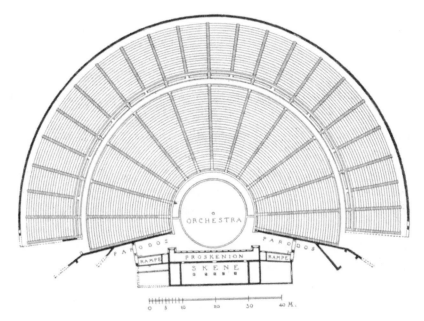

Figure 12: Plan of the theater at Epidaurus. From Allan Marquand, Greek Architecture *(New York: Macmillan, 1909).*

It is also possible that the theater's acoustics were enhanced, as Declercq has suggested, by the "periodicity" of the seats, that is, by the deliberate spacing of the rows to conform to mathematical principles that underlay harmony as the Greeks understood it. By experimenting before construction with the acoustical effects of various seat sizes, shapes, and placements on the hill's slopes, Polycleitus the Younger could have perfected his design. As Vitruvius would later write: "Following in nature's footsteps, the architects of old designed the slopes of their theaters by investigating how the voice travels. Applying principles of mathematics and music, they attempted to make voices from the stage reach the ears of the audience more clearly and sweetly. ... Thus from antiquity, theaters were planned in accordance with the principles of harmonics in order to amplify the human voice."[3]

Most significantly at Epidaurus, the Greek science of optics worked hand in hand with the Greek science of acoustics to produce an extraordinary building that not only gave spectators a flawless view but also enabled them to hear music and the human voice as never before.

ACOUSTICAL POWER

Though ours has often been labeled an age of visual imagery because of the pervasive influence of print advertising, movies, and television, it can with equal justification be called an auditory age. Thanks to the creative contributions of contemporary acoustical engineers, we surround ourselves in concert halls, multiplexes, and home theaters with pleasurable sounds of the highest fidelity. We cruise in soundproofed cars, bathed in our favorite melodies or attuned to the reassuring words of our favorite authors. With the intimate aid of earbuds we can even insulate ourselves in a comforting sonic blanket from the world outside, or converse by cell phone with far-off family and friends as though they were walking right beside us. The curious scientists of ancient Greece could hardly have imagined what electronic marvels their early explorations of sound would someday inspire.

Chapter 6

MECHANICS

Two qualities separated the ancient Greeks from their anthropomorphic gods: immortality and superhuman power. For both these reasons, the Greeks worshipped and envied them.

Yearning to attain divine status but knowing they must inevitably die, the Greeks defied death by performing legendary acts of bravado in battle and by fashioning undying works of literature and art. "To the extent we can, we must strive to be immortal," said Aristotle.[1] Yet however much their martial and aesthetic feats launched their names onto the boundless ocean of fame, the all-too-human Greeks were like puny mariners on a fragile ship caught out on the open sea before the gale force of the gods' implacable wrath.

Their relative weakness, however, did not prevent them from celebrating those powers they did possess. Every four years at the Olympic Games, they proudly offered up to Zeus exemplars of their strength, their stamina, and their speed. Their myths, moreover, were populated by half-breed heroes like Achilles, mothered by the sea goddess Thetis, or Hercules, fathered by Olympian Zeus himself—musclemen who showed what memorable things a man could achieve if divine blood flowed through his veins (fig. 13).

Conversely, the thing they feared the most was impotence. Thus, the three arch sinners who were singled out for extraordinary punishment in Hades were tortured by the frustration of failing to physically accomplish their goals: Tantalus, forever unable to scoop up water to slake his thirst or pick fruit to salve his hunger; Tityus, unable to beat off the ravenous vultures that constantly tore at his liver; and Sisyphus, again and again denied the ability to shove a huge rock over the crest of a hill.

Figure 13: Hercules leaning on a club draped with the skin of a lion he had slain. Naples Museum. From Franz von Reber, History of Ancient Art, rev. ed., trans. and augm. Joseph Thacher Clarke (New York: Harper & Brothers, 1882).

Bracing himself with his hands and feet,
he kept pushing the stone up the hill.
But when he was just about to shove it over the crest,
its overpowering weight drove it back,
and once again down to the plain below it rolled relentlessly.
Then straining, he gave it another push, the sweat
dripping from his limbs, the dust rising above his head.[2]

 Given their intellectual curiosity, their perennial fascination with power, and their compulsion to test intrinsic limits, it is no wonder that the Greeks founded the science of mechanics, a science that investigates the dynamic interplay between matter and energy. Indeed, even the Olympic Games could inspire their thinkers to explore the impact humans could have on the material world.

ARISTOTLE AT THE OLYMPICS

Three of the greatest philosophers of Greece were fans of the quadrennial events that took place at Olympia: Socrates was so dedicated he walked for five or six days from Athens just to get there. Plato trained until he became a prize-winning wrestler. And Aristotle so admired the beauty of athletic physiques that he kept a faithful record of all the Olympic champions.

But to a philosopher like Aristotle, there was more to sports than might meet the average eye, just as to a scientist like Sir Isaac Newton, there was more than meets the eye in seeing an ordinary apple fall from a tree.

Newton's assistant, John Conduitt, recalled the epoch-making incident:

> In the year 1666 . . . while he was pensively meandering in a garden it came into his thought that the power of gravity (which brought an apple from a tree to the ground) was not limited to a certain distance from the earth, but that this power must extend much further than was usually thought. Why not as high as the Moon he said to himself and if so, that must influence her motion and perhaps retain her in her orbit, whereupon he fell a calculating what would be the effect of that supposition.[3]

In the same way, though we cannot prove it, watching the athletic competitions at Olympia may have led Aristotle and others of like mind to speculate on the invisible physical principles that underlay what they saw.

Watching athletes compete in footraces may have led him to ponder the correlation between a runner's speed, the distance he covers, and the time it takes him to do it—and then apply that same thinking to conceptualizing the relationship between an object of a given weight, on the one hand, and, on the other, the force needed to move it a given distance.

Aristotle wrote:

> If a given force is moving something, it stands to reason it has already been moving it within something and over something. (By "*within* something," I mean within a period of time; by "over something," I mean over a distance. For

if something is producing motion, it stands to reason it has already been producing that motion over a given distance and within a given period of time.) Consequently, if force A has moved object B a distance of C within a time of D, it follows that, within that same period of time D, that same force A would move ½ B twice the distance C; likewise, within a time of ½ D it would move ½ B the whole distance C: for their relationship would be analogous.[4]

The main thing to observe here is how the Greek mind extracted from real-life experience a principle that had potentially universal applicability and then translated it into abstract mathematical terms that could be applied to other phenomena. In addition, we see how the Greek mind applied to its interpretation of the universe the very same sensitivity to proportionality that characterized its reflections on such things as the conduct of one's life (the role of reason vs. the role of emotion), the dynamics of politics (the needs of the individual vs. the demands of the state), and the creation of art (the relative sizes of components in architecture and of anatomical parts in sculpture).

Figure 14: An ancient marble copy of Myron's Discus Thrower. *From Max Sauerlandt,* Griechische Bildwerke *(Düsseldorf and Leipzig: Karl Robert Langewiesche, 1907).*

Similarly, watching a contest of discus throwing (fig. 14) or javelin throwing may have led Aristotle to question why an object flies through the air even after it has left the hand of the thrower and then suddenly stops.

Aristotle pondered:

> Objects that are thrown move even when the thing that caused them to move is no longer touching them. This happens either because of displacement, as some say, or because the air that has been accelerated by impact drives the projectile faster than it would otherwise go. . . .
>
> No one can say why something set into motion stops somewhere. Why *here* instead of *there?* You would think a thing would either stay at rest or, once moved, would keep on moving unless something more powerful got in its way.[5]

In reaching here toward the discovery of the principle of inertia, Aristotle was not afraid to ask questions even when he had no definitive answers. Such implacable curiosity is, in fact, at the very heart of scientific discovery, and for the Greeks, that curiosity came naturally and was joined to a fierce love of freedom.

Not far from the stadium at Olympia was the temple of Zeus, the god whom the festival honored. Within the temple was one of the legendary Seven Wonders of the World, the immense gold and ivory statue of Zeus, seated on his Olympian throne (fig. 15).

Figure 15: Coin portraying Zeus enthroned in his temple at Olympia. From George Redford, A Manual of Sculpture *(London: Sampson Low, Marston, Searle, and Rivington, 1882).*

Greek temples like this and like the Parthenon in Athens were built of walls made of marble blocks and lofty columns made of marble drums piled one upon the other. To hoist the blocks and column drums in place and slide them into position, laborers used pulleys and levers, relatively simple mechanical devices that Aristotle and his contemporaries would surely have been familiar with. But the science behind these mechanical contrivances was another matter altogether. Exactly how *did* they make work easier? What allowed them to multiply the force of the human muscular system so that an ordinary man became a superman?

While the physical principle behind the pulley may not have been fully understood until Hellenistic times, the principle behind the lever seems to have been understood in Aristotle's day or a century or so later. The mathematical formula explaining its operation can be found in a work ascribed to Aristotle called *Mechanics*. In it we read that a weight and the length of a lever needed to move that weight are inversely proportionate to each other. The heavier the object, the longer the lever must be. Put another way, the closer the fulcrum is to the object to be lifted, the easier the lifting becomes. While this fact may have been familiar to stonemasons, it was scientists who first discovered the mathematical ratio that underlies it—a universal "law of the lever" that governs the interaction of matter and energy. In a later century, the Hellenistic engineer and inventor Archimedes, recognizing the theoretically unlimited power of leverage, would make a cosmically grandiose but fundamentally valid boast: "Give me a place to stand and I will move the Earth."[6]

Of course, the law of the lever was also used to explain tasks that were much more mundane: why, for example, oarsmen sitting amidships on a boat and manning longer oars have an easier time pulling against the sea (answer: because the physical effort diminishes the farther an oar extends toward the rower from the fulcrum, or thole-pin); or why a pair of pliers, or forceps, makes it easier for a dentist to extract a tooth (answer: because the forceps is really a set of levers joined together).[7]

The Engines of War

It wasn't long before the practical advantages of such mechanical principles were applied to the art of war. Earlier we saw how Archimedes may have used his insights into the science of optics to design a "death ray" that could be aimed at Roman ships during the siege of Syracuse in 213 BCE. Though he regarded the practical application of his knowledge as demeaning,[8] Archimedes in time of war was persuaded to use his engineering genius and expertise with levers and counterweights to turn a giant mechanical "claw" into a devastating weapon.

> The ships that entered closer in to the harbor to be safely inside the range of the catapults were attacked by an iron claw attached by a strong chain to a swing-beam that projected out from the city wall. When the claw was thrown onto a ship's bow and a heavy lead counterweight dropped to the ground, the device stood the ship up on its stern with its bow suspended. Then, with the sailors in a panic, the device suddenly let the ship go, causing it to tumble from the heights of the wall and splash into the sea so that, even if it landed right-side up, it took on a huge amount of water.[9]
>
> Often a ship, lifted out of the sea, was whirled around and dangled in mid-air—a sight to make one shudder—while its crew was tossed out and scattered willy-nilly, until the ship was dropped empty onto the walls or slipped out of the loosening claw's grip.[10]

During the warring centuries of the Hellenistic period, military engineering became one of the most salient aspects of applied science. Its special focus was siege-craft, and its practitioners included Archimedes's fourth-century BCE predecessor Aeneas Tacticus (the Tactician); Archimedes's third-century BCE contemporaries, Biton of Pergamum and Ctesibius; and his successors, Philon of Byzantium (fl. 200 BCE), Athenaeus Mechanicus (the Mechanical Engineer; first century BCE), and Heron of Alexandria (first century CE). Many of their handbooks still survive. Among their innovations were the mobile scaling ladder, the rapid-fire arrow launcher, the variable-range catapult, the spring-loaded catapult, and another type of catapult that was powered by compressed air (fig. 16).

In our own day, when armaments have been promoted by a military-industrial complex as the surest guarantor of peace, the following naively optimistic remarks by Heron of Alexandria seem eerily prescient:

> The largest and most essential part of philosophy deals with tranquility, to which a great deal of thought has been devoted and is still being devoted by those who concern themselves with learning. I for my part, however, think

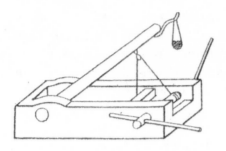

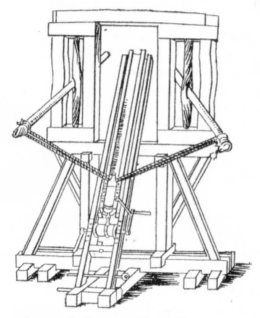

Figure 16: Examples of ancient catapults. From Harry Thurston Peck, ed., Harper's Dictionary of Classical Literature and Antiquities *(New York: American Book Company, 1896).*

the search for tranquility will never reach a definitive conclusion by bandying words.

Mechanics, however, has already taught mankind how to live a tranquil life through its actions, which surpass the words of any teacher—specifically through that branch of mechanics known as artillery construction.

With its aid men will never need fear the attacks of enemies at home or abroad, in peacetime or in war, if for every time and on every occasion adequate provisions are made for the manufacture of military machinery. . . . And since those who know siege-craft also know how to defend against it, they will always be able to live without fear, while potential aggressors, seeing their enemy's level of preparedness, will not dare to attack.[11]

A Moving Experience

The inventiveness of the Greek mind found expression not only on the battlefield but in the world of entertainment. In the previous chapter we saw how the science of acoustics enhanced theatrical performances. The science of mechanics performed a similar dramatic role.

To display onstage events that had just occurred offstage (the suicide of a character, for example), the Greek theater used a device called an *ekkuklema*, which literally means "something wheeled out." Although we don't know exactly how such a device may have operated, there are two possibilities. Either it was a kind of platform that could be wheeled out onstage through the main door of the set, carrying its tableau, or it was in fact a revolving stage that, together with the "door," could be rotated to reveal an interior scene.

Another device frequently used in the classical Greek theater was the *mechane* (may-hah-NAY), or "machine." The *mechane* was a crane positioned behind the stage that could be raised with pulleys and swung out high over the stage. Dangling from it in a basket or harness would be an actor playing the role of a god or goddess making his or her appearance and pronouncements from heaven to save the day for the other characters and make everything come out right in the final act. In Latin, such a figure was referred to as a "deus ex machina," or "a god from a machine," an expression we still use today to refer to a person or thing whose sudden and unexpected appearance has game-changing results.

Just down the street from Athens' Theater of Dionysus was another machine devoted to drama: the drama of Athenian democracy. Standing in the Agora—Athens' public square—was a slot machine called a *kleroterion* that was used to pick jurors. It was a thick slab of stone measuring about three feet wide by five feet tall and cut with as many as 550 slots arranged in regular columns and rows. A prospective juror would approach the device and hand a special wooden or bronze identity token to the presiding magistrate, who would insert it into one of the open slots in the machine. After all the slots were full, the magistrate would pour a bag of black and white balls into a metal funnel attached to the top of a metal tube running down the side of the machine. As a crank at the bottom was turned, a ball would drop out (not unlike the workings of a modern gum ball dispenser). If the ball was white, all the prospective jurors in a given row would be impaneled to serve on a jury. If the ball was black (or perhaps vice versa), the jurors in that row would be excused. Then the magistrate would turn the crank again to pick another set of jurors for a different trial.

An original but broken *kleroterion* is on exhibit today as one of the prized possessions of the Agora Museum, prized not merely because it is a rare artifact of classical Athens but also because it symbolizes one of the greatest gifts of Greece to the modern world: the jury system, a Hellenic invention that testifies to the humanistic faith in the power of individuals as a group to secure justice through the exercise of their combined intelligence.

THE MUSEUM OF MARVELOUS MECHANICAL CONTRAPTIONS

The ingenuity of the Greek people blossomed during the Hellenistic age, turning mythical imaginings into practical realities. This was strikingly true in the fields of automation and robotics.

The archetype of the Greek inventor was Daedalus, who in legend is said to have lived centuries before the Trojan War. According to the story, he prospered in Athens but fled to the island of Crete to escape prosecution for the murder of a professional rival. He soon became royal architect to King Minos of Crete, designing for him the labyrinthine prison that would hold captive the half-man, half-bull monster known as the Minotaur. Imprisoned by Minos for giving away the secret of the prison's design, Daedalus escaped from the palace with his son Icarus by constructing makeshift pairs of wings from feathers, wax, and wood, thereby becoming (along with his son) the first man in history to fly and the mythical founder of aeronautical science (fig. 17).

Legend has it that Daedalus also fashioned statues that were so realistic they could see and walk.[12] In addition, it is said he carved a wooden statue of the goddess Aphrodite that could move when he poured mercury into it.[13]

Another resident of Crete, a fabulous android called Talus, was made of bronze and is said to have guarded the island by walking sentry duty along its coastline three times a day, hurling rocks at any invaders he came upon,[14] or killing them by turning red hot and burning them to death in his molten embrace.[15] Talus's only point of vulnerability, Apollonius tells us, was a vein that transported his "blood."

The oldest-surviving work of Greek literature, Homer's *Iliad*, tells how,

Figure 17: Daedalus constructing wings for himself and his son Icarus. From an ancient bas-relief at the Villa Albani, Rome. From Harry Thurston Peck, ed., Harper's Dictionary of Classical Literature and Antiquities (New York: American Book Company, 1896).

in the days of the Trojan War, the craftsman-god Hephaestus made serving trays and equipped their stands with golden wheels so they could travel automatically between his workshop and the place where the gods partied. Moreover, to help him move about in his smithy, the lame god is said to have fashioned maidservants out of gold who could think, speak, and move.[16]

Their imaginations stirred by such tales, the Greeks endeavored to make real-life robots, or at least the semblance of same. In the course of discussing the difference between actual and potential motion, Aristotle became the first Greek author to allude to "automatic marvels" he had seen with his own eyes, marvels that seemed to move on their own, but alas he gave no real details about *how* they moved.[17] Elsewhere, he describes artificial frogs he had seen submerged in water, which then rose to the surface one after the other as the salt that weighed them down dissolved.[18]

Perhaps the greatest single masterpiece of ancient automation was devised by the Hellenistic genius Heron of Alexandria. It was an amazing "automatic theater." He described its design in great detail, a design that included a motor driven by the pull of a slowly descending weight, the world's first rotating camshaft, and scenic action that occurred three-dimensionally in multiple planes. Heron outlined one of many possible scenarios that could be presented on its stage:

> Initially, when the automaton is positioned at its starting point and we step back, after a short time it will begin moving toward the finish line. When it stops, the altar of Dionysus will ignite, milk or water will squirt from the god's scepter, and wine will pour from his cup onto the panther at his feet. Each of the areas that adjoin the four columns of the base will then be covered with a garland. Next, the god's devotees, the Maenads, will go dancing in a circle around the shrine. The sound of drums and cymbals will be heard. Afterwards, when the noise has stopped, Dionysus will turn away, while simultaneously the goddess of Victory on the roof will rotate. When the altar comes in front of Dionysus again, it will once again ignite, and liquid will again squirt from his scepter as will wine pour from his cup. The Maenads will again dance in a circle around the shrine to the accompaniment of drums and cymbals. And when they stop again, the automaton will return to its original location. . . .
>
> At the bottom of the base where the wheels are located, there is a container holding lead balls that can roll out onto the base's floor. From there they can easily drop through a hole of sufficient size which can be opened by a door operated by a cable. Under the hole is a drum attached to a cymbal. As the balls drop through the hole they first hit the drum and then drop onto the cymbal to complete the sound effect.[19]

By adjusting the pegs on its spindles and the lengths of the cables that operated the various components, the mechanical theater could be reprogrammed—with a change of scenic backdrops and wooden "actors"—to produce a whole variety of surprising new movements and dramatic effects.

One of Heron's "plays" was based on the tale of Nauplius, a king who determined to wreak vengeance on the Greeks for killing his son during the Trojan War. He did so, according to the legend, by luring their ships onto dangerous rocks as they sailed back to Greece. In a programmed sequence of scenes, Heron's theater pictured the Greek ships putting out to sea one by one; dolphins diving over the waves; the sea growing stormy; Nauplius signaling to the fleet by torchlight and trumpet; the vengeful goddess Athena lending him aid; the Greek ships foundering; and the hero Ajax swimming among the waves until he is struck by lightning and drowns.

Heron also constructed another device with a hidden stopper that could pour out wine a cupful at a time, and he invented a self-refilling serving bowl for wine that replenished itself automatically, thanks to an unseen flotation device (not unlike the mechanism inside the tank of a modern flush toilet that refills the toilet bowl up to a predetermined level).

Heron's other inventions included a temple with double doors that magically opened to the sound of a trumpet and closed by themselves when a fire was lit at the temple's altar (fig. 18). The secret of the doors' operation was hidden in the sanctuary's basement. Heated by fire, the air inside a sealed underground water tank expanded and forced water inside the tank to flow into a suspended bucket. As the bucket grew heavier and sank, it pulled two chains wrapped around belowground extensions of the doorposts. As the doorposts rotated on their pivots, they opened the temple doors. When the fire on the altar later went out, the air trapped inside the sealed tank condensed, siphoning water from the bucket back into the tank. As the bucket in turn grew lighter, it rose with the help of a pulley and counterweight, making the underground doorposts rotate in the opposite direction and slowly close the doors above. At the start of the cycle, air pressure from the heated tank could also be diverted to blow a trumpet as the doors of the temple miraculously opened.

With the aid of compressed air, Heron also devised a metal tree with metal birds perched on it that "sang," and a fountain triggered by air pressure that operated when a sealed container was heated by the rays of the sun, expanding the air trapped inside. Indeed, it was Heron's understanding of the essential nature of air that allowed him to invent such marvels. Rather than regarding air as mere emptiness, Heron asserted that air was a substance that occupied space. He offered experimental proof for his assertion:

> Vessels which seem to most men empty are not in fact empty, but full of air. Now air, as those who have studied physics agree, is composed of particles minute and light, and for the most part invisible. If, then, we pour water into an apparently empty vessel, air will leave the vessel in proportion to the quantity of water which enters it. This may be seen from the following experiment. Invert a vessel which seems to be empty and, while carefully

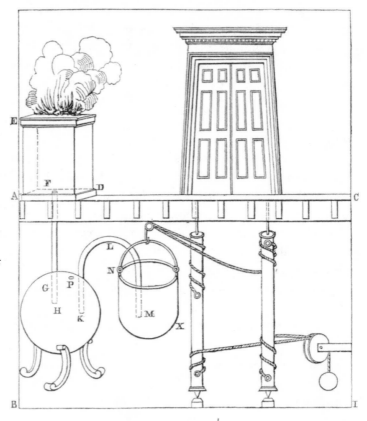

Figure 18: The design for an automatic temple door opener. From Bennett Woodcroft, ed., The Pneumatics of Hero of Alexandria, trans. Joseph Gouge Greenwood (London: Taylor, Walton, and Maberly, 1851).

keeping it upright, press it down into water. The water will not enter it even though it is entirely immersed. This shows that the air, being matter, and having itself filled up all the space in the vessel, will not allow the water to enter. Now, if we bore a hole in the bottom of the vessel, the water will enter through the vessel's mouth while the air escapes through the hole. Again, if, before perforating the bottom, we raise the vessel vertically, and turn it up, we shall find its inner surface entirely free of moisture, exactly as it was before immersion. Hence, it must be assumed that air is matter. The air when set in motion becomes wind (for wind is nothing else but air in motion), and if, when the bottom of the vessel has been pierced and water is entering, we place our hand over the hole, we shall feel the wind escaping from the vessel. This is nothing other than the air which is being driven out by the water. We must not then suppose that a vacuum generally exists in nature.[20]

The realization that if air is eliminated from an enclosed space another substance would tend to fill it inspired Heron or one of his predecessors to invent the siphon. That realization along with another—that air expands when heated and contracts when cooled—led to a variety of ingenious applications, some of which have already been described. One of the simplest applications, yet one of the most beneficial to humanity, has been the hypodermic syringe. Here is Heron's detailed description of its ancient construction.

> A hollow tube, of some length, is made, AB; into this another tube, CD, is nicely fitted to the extremity C of which is fastened a small plate or piston, and at D is a handle, EF. Cover the orifice A of the tube AB with a plate in which an extremely fine tube, GH, is fixed, its bore communicating with AB through the plate. When we desire to draw forth any pus we must apply the extreme orifice of the small tube, H, to the part in which the matter is, and draw the tube CD outwards by means of the handle. As a vacuum is thus produced in AB something else must enter to fill it, and as there is no other passage but through the mouth of the small tube, we shall of necessity draw up through this any fluid that may be near. Again, when we wish to inject any liquid, we place it in the tube AB, and, taking hold of EF, depress the tube CD, and force down the liquid until we think the injection is effected.[21]

Modern applications of this astoundingly simple principle include a piston-driven device unknown to the ancient Greeks—the internal combustion engine—that has revolutionized modern transportation. Another related device is the piston-driven force pump, originally designed by Heron to fight fires in Alexandria by using twin cylinders to generate high-pressure jets of water that could be directed at burning buildings.

Less dramatic but no less effective for pumping purposes was a unique pump attributed, by ancient Greek and Roman sources, to the legendary Archimedes. Known as the "Archimedes screw," it consisted of a large screw-shaped shaft equipped with a handle at one end. The screw was fitted snugly into a long pipe. As the pipe was inserted into water and the handle was turned, the water was scooped up by the turning screw and carried upward to the surface on the screw's blades (fig. 19). The faster the handle was turned, the more water was brought up. Archimedes's screw was employed for irrigation (to raise river water to the level of the adjacent fields), for emptying bilge water out of the bottoms of boats, and even for moving grain out of deep storage. In terms of its basic mechanics, it was nothing more than an inclined plane wrapped in spiral fashion around a cylinder—an engineering solution of astounding simplicity.

Two of Heron's predecessors, Ctesibius and Philon of Byzantium, devised other noteworthy mechanical marvels. Ctesibius created a self-regulating water clock as well as an organ that automatically played music with the help of water pressure. Philon, for his part, designed a coin-operated vending machine for temples that, at the drop of a coin, dispensed holy water and balls of pumice soap to worshippers who wanted to cleanse themselves before praying.

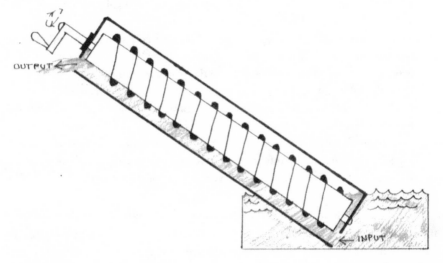

Figure 19: Archimedes's screw pump. Drawn by the author.

As T. E. Rihll has pointed out, "Some scholars have thought some of the machines described in their surviving texts to be fanciful ["Rube Goldberg"] designs. But these scholars ignore other ancient literature which takes for granted the existence of such machines as real, working, and familiar to their audience."[22]

While some of these devices had practical utility, others were created chiefly to demonstrate in virtuoso fashion the ingenuity of the human intellect whose display could entertain and delight the Greek mind as could nothing else. To be sure, the success of such intellectual ventures required an audience as enthralled by such creative marvels as were the men who had first invented them, men who, in setting an artificial universe into motion, became as gods.

Two other Greek inventions are deserving of mention. One, the so-called Antikythera mechanism, is the world's oldest computer (to be discussed in chapter 10). The other is a "steam engine" devised by Heron (fig. 20).

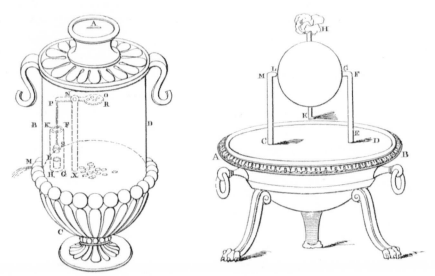

Figure 20: A coin-operated holy water dispenser (left) and a model of Heron's steam engine (right). From Bennett Woodcroft, ed., The Pneumatics of Hero of Alexandria, *trans. Joseph Gouge Greenwood (London: Taylor, Walton, and Maberly, 1851).*

Heron's mechanism consisted of a glass ball with two vertical jets spaced halfway apart that pointed in the same direction. Filled with water, mounted on a horizontal axle, and then heated from below, the device produced steam that vented from the ends of its bent jets, spinning the ball around in obedience to Newton's third law of motion with a velocity perhaps greater than that of any other machine in the ancient world.

In his illustrated manual *Pneumatics*, Heron gave straightforward instructions for reproducing this remarkable invention:

Place a cauldron over a fire: a ball shall revolve on a pivot. A fire is lighted under a cauldron, AB, containing water, and covered at the mouth by lid CD; with this the bent tube EFG communicates, the extremity of the tube being fitted into a hollow ball, HK. Opposite to the extremity G place a pivot, LM, resting on the lid CD; and let the ball contain two bent pipes, communicating with it at the opposite extremities of a diameter, and bent in opposite directions, the bends being at right angles and across the lines FG, LM. As the cauldron gets hot it will be found that the steam, entering the ball through EFG, passes out through the bent tubes towards the lid, and causes the ball to revolve.[23]

It seems Heron could not imagine how this wondrous invention could be put to use. Had he done so, he might have set the Industrial Revolution itself into motion more than a millennium and a half before its historic inception (except for the fact that the emphasis on craftsmanship and the ample supply of slaves in the Hellenistic world undercut the need for machine production).

But the absence of a practical use for his invention did not diminish Heron's exhilaration when first his engine spun. The thrill he felt was no less than that experienced by Greek discus throwers whose striving served no higher purpose than to celebrate the pursuit of excellence and man's glorious triumph in attaining it.

THE POWER OF ATTRACTION

Fascinated as they were by machines that seemed to move by themselves, the Greeks were perhaps even more intrigued by objects that possessed the seemingly magical power to make other objects move without any human intervention. The mysterious objects in question were pieces of amber and lodestones.

Apart from its use in jewelry, the fossilized resin we call amber can, when rubbed, attract lightweight materials to it because the friction caused by rubbing creates static electricity. The Greeks, however, did not understand the workings of static electricity. They assumed there was some mysterious power within the amber itself that allowed it to act as though it were conscious and alive. Taking its cue from the warm, sunlike glow of the substance, Greek mythology explained that pieces of amber were in fact tears shed by the sun-god's daughters. According to the story, the daughters mourned the death of a young man named Phaëthon, who had been killed while trying to drive the chariot of the sun. As their tears fell to the ground, they were hardened by the heat of the sun and became bits of amber. The Greek name for amber, *electron*, is in fact derived from the Greek word for the sun, *helios*. And from the ancient word *electron* comes our modern word *electricity*.

It was the early Greek philosopher Thales (seventh to sixth century BCE) who first proposed that amber possessed a *psyche*, or "soul," that made it act like an animate being. Two centuries later, the philosopher Empedocles (ca. 492–43 BCE) advanced the theory that there were atomic effluences that flowed outward from amber and filled empty pores in the surfaces of other materials, thereby attracting them. To Thales and Empedocles, the same fact also explained how lodestones magnetically attracted pieces of iron.

The word *magnet* is also derived from Greek. According to one ancient source, the word echoes the name of a legendary shepherd named Magnes, who became the first man to find a lodestone when one clung to the iron nails in the sole of his shoe. A second etymology says the word recalls a land named Magnesia, either the one near Macedonia or another of the same name located in Turkey, where lodestones were commonly found. These magnets were also called "Heraclean stones" by the Greeks, either because they possessed supernatural power like the mythical hero Hercules or because so many were found in Hercules's namesake city of Heracleia in Asia Minor.

Besides the philosophers Thales and Empedocles, a wide array of Greek thinkers across the centuries speculated on the nature of magnetism. The fifth-century BCE philosopher Democritus believed that whirling atoms emanating from a lodestone and a piece of iron intermingled and brought the two objects together. Discounting such atomic theories, the second- to third-century CE physician Galen argued for an innate power of attraction, similar to the way the right medicine can draw out poison from a wound. Despite their persistent interest in magnetism, however, the Greeks did not understand a magnet's polarity or its attraction to Earth's magnetic pole, nor did they, as far as we know, create magnets artificially.

By Hellenistic times, the magic of magnetism was being put into the service of religion. In the third century BCE, the Alexandrian architect Timochares designed, but did not live to erect, a temple dedicated to King Ptolemy and his wife Arsinoë, in which an iron figurine of the queen was to be suspended in midair using the power of opposing lodestones. Also, according to the fourth- to fifth-century CE poet Claudian, a magnetic figurine of the goddess of love, Aphrodite, and an iron figurine of her divine lover, Ares, were used in rites at her temple in Alexandria, rushing of their own accord into each other's arms as a priest held the two figurines in proximity.

Besides the occasional use of magnetism for religious purposes, Greek authors also used it as a humanistic metaphor. For example, the fifth- to fourth-century BCE playwright Eubolus and the second-century CE novelist

Achilles Tatius used it as a symbol for the erotic attraction each had experienced; while Plato (*Ion*, 533 D–E) let it symbolize the inspirational power the written word can have upon a performer, and the power an inspired performer can in turn have upon his rapt audience.

Chapter 7

CHEMISTRY

L ike divers plunging through the shimmering swell to harvest sponges from the shadowy floor of the sea, Greek thinkers yearned to get to the bottom of things—to penetrate the glittering surface of appearance in order to retrieve the unseen essence that lay hidden in its depths. To their minds, no more fundamental essence existed than the elements out of which physical reality itself was fashioned, and it is this essence that they sought to define.

The question "What is the world made of?" might seem impractically abstract, given life's pressing demands. Yet, as Socrates once observed, "A life that can't be questioned isn't worth living."[1] When Socrates uttered these words, he was indicting a political system intent on crushing its enemies, but the words apply equally well to the cause of Greek science, for to the ancient Greeks the quest for ultimate knowledge was akin to breathing.

THE PHILOSOPHER-SCIENTISTS

The first Greek scientists were philosophers, literally "lovers of wisdom," who speculated on the nature of the universe. Though their speculations were, in fact, exercises in theoretical physics, the issues they debated became central to the science of chemistry.

Collectively, these early scientists are called "pre-Socratic" philosophers because most of them were born before Socrates (469–399 BCE). Some were "monists" (from the Greek word *monos*, "alone"), philosophers who believed that all substances share one and only one fundamental ingredient, such as water, air, or fire—or, according to Anaximander, a mysterious and imperishable element called "the infinite," from which all things arise.

Others were "pluralists" who intellectually championed a plurality of common ingredients. For Empedocles, it was earth, water, air, and fire, which he called reality's four "roots." In symbolic terms, what Empedocles meant was that reality is either solid (earth), liquid (water), gas (air), or energy (fire), or a mixture of some or all of these properties held together by attraction until their material dissolution, with the proportion of each ingredient determining the peculiar characteristics of any given compound. Anaxagoras would "improve on" Empedocles's cosmic recipe by adding yet another ingredient, mind or intelligence, while their contemporary Democritus would argue instead that at its most fundamental level matter consists of a multiplicity of indivisible particles called "atoms" (from the Greek word *atomos*, "indivisible"). Plato, Socrates's student, would go on to describe in geometric terms the distinctive molecular structure of each of Empedocles's roots, for the first time naming them "elements" (*stoicheia*), and adding a fifth celestial element he called "ether."

To be sure, little of this sounds like science to the modern ear because it is pure speculation, undocumented by empirical evidence and untested by experimentation. But it is also a long way from a purely mythic account of the universe, like the one offered by the seventh-century BCE poet Hesiod ("In the beginning was Chaos, and from it arose broad-bosomed Earth, the trusty foundation of the immortal gods who together inhabit the peaks of snowy Olympus")—an account that focuses not on the actual substance of the material world but on the supernatural realm of the gods.

Nevertheless, we should not assume that early Greek science was religion-free. Empedocles, for example, who wrote in inspired verse and even viewed himself as a prophet, assigned to his four roots the names of deities and attributed the forces of attraction and dissolution to divine powers. And Pythagoras, another leading pre-Socratic, founded a religious cult based on the mystical belief in cosmically significant numbers. Thus, however rational Greek science might seem in its quest for truth, it did not entirely divorce itself from spirituality.

Religious considerations aside, the early philosopher-scientists of Greece were instrumental in laying the groundwork for modern science and, in particular, the science of chemistry: first, because they made a determined intellectual effort to understand the nature of the physical world; second, because they proposed that matter consisted of specific elements and fixed combinations of elements; and third, because they introduced and later elaborated upon an atomic theory. Though modern chemistry's periodic table of elements (now standing at 112 confirmed elements and counting) far exceeds in number the four "roots" of Empedocles, and though Democritus's indivisible "atom" was long ago split by physicists, the theorists of ancient Greece blazed the path to our current understanding of the material world.

A Mediterranean Think Tank

Despite belonging to schools of thought (the monists and the pluralists, for example) and despite often knowing and at times challenging one another's theories, the pre-Socratic philosophers of the seventh, sixth, and fifth centuries BCE were largely independent thinkers who, except for sharing their ideas with their disciples, deliberated alone. Indeed, many lived at the far-flung edges of the Hellenic world, from the coastal cities of western Turkey to the colonies of southern Italy and Sicily where Greeks had settled in the past. Some may have even traveled to the lands of the Near East, to Egypt and Babylonia, where they would have imbibed the more ancient learning and wisdom of those earlier civilizations. With the fall of classical Athens in 404 BCE, however, and with the eventual rise of a dynamic postclassical, or Hellenistic, culture, a radical change took place that brought great Greek thinkers together within the confines of a single city—indeed, within the walls of a single building.

The city was Alexandria, Egypt, founded in the fourth century BCE by its namesake, Alexander the Great, as the cultural capital of his three-thousand-mile-wide empire (fig. 21). The building was a center for advanced study established by Alexander's successor, King Ptolemy I of Egypt. Named for the Muses, the patron goddesses of the arts, the research center was appropriately dubbed the "Museum." Aided by a vast library housing the wisdom of the known world, the community of scholars that resided there constituted a critical mass of immense intellectual talent.

The port city of Alexandria itself, thriving on Mediterranean trade, became a magnet for immigrants from Greece and the Levant, thereby fulfilling Alexander's dream of a global multicultural society that would draw its creative strength from the energizing intermingling of diverse ethnic ideas and outlooks. Different religious groups, both Jewish and Christian, eventually flocked to Alexandria and rubbed shoulders with the worshippers of native gods such as Isis and Osiris. It was within this stimulating ideological and spiritual milieu that Greek scientific inquiry gained new impetus.

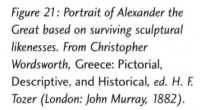

Figure 21: Portrait of Alexander the Great based on surviving sculptural likenesses. From Christopher Wordsworth, Greece: Pictorial, Descriptive, and Historical, *ed. H. F. Tozer (London: John Murray, 1882).*

LABORATORIES IN THE DESERT

When the seeds of Hellenic thought fell on the fertile soil of Egypt, they were dispersed on a soil already rich in technological expertise, especially in the field of applied chemistry. For over three millennia, the metallurgists of ancient Egypt had perfected the process of smelting copper, the pre–Iron Age metal of choice for weapons and tools, extracting it from its ore by heating it over charcoal (at times with the addition of a silicate or iron oxide flux to reduce the slag's melting temperature) or extracting it from malachite (by employing blowing tubes to generate higher temperatures). Discovering that copper could be strengthened by alloying it with another mineral, Egyptian workers blended it, at great personal risk, with locally available arsenic, or, more safely, with harder-to-find tin. At the same time, artisans discovered a way to color gold, the preferred metal of Egyptian royalty, by tinting it with iron oxide to give it a reddish cast.

According to recent research, advanced chemistry in Egypt also played an important role in the development of milady's cosmetics. While blue pigments were produced with the help of fire, some green, white, and black pigments were manufactured by using "wet" chemistry to create supplies of synthetic laurionite and phosgenite, minerals very rare in their natural state. When applied as eyeliners, these lead-based compounds chemically stimulated the wearer's immune system, inhibiting the growth of bacteria that could infect the eyes.[2]

Egyptian craftsmen also excelled at making glass and endowing it with brilliant colors that simulated those of precious gems.

Universally acknowledged by the Greeks as the world's most adept healers, the physicians of Egypt had a panoply of drugs at their disposal with a range so wide it was praised by the poet Homer and, in later centuries, by the Roman scholar Pliny. For those patients who didn't recover, a compound of sodium called natron was applied to their corpses in crystalline form. Doing triple duty as a powerful degreaser, dehydrating agent, and disinfectant, natron guaranteed spiritual immortality by protecting dead bodies, the vessels of the spirit, from decay. As mummies, these embalmed corpses still testify to the chemical expertise of the Egyptians.

FROM SCIENCE TO ALCHEMY

Philosophical and religious currents eventually began to flow in the land of the Nile that would determine the future direction of chemical science.

The end of Greece's classical era and the emergence of an intellectually expansive Hellenistic world had strained the bonds between the Greeks and the old gods of Olympus. New ideas were in the air, replacing traditional rituals and beliefs with novel, personal approaches to the search for spiritual fulfillment. Epicureanism challenged the very notion of the gods ruling an orderly universe, arguing instead that in a random universe it was the individual's obligation to seek his or her own pleasure. Stoicism, for its part, emphasized personal pain and the need for people to transcend it through philosophical discipline. The two themes, sensualism and struggle, also pervaded Greek sculpture, as did individualism itself through the emergence of realistic portraiture as a major genre of art. By the third century CE, another philosophy, Neoplatonism, appeared. Expanding upon Plato's theory that reality as we know it is an imperfect replica of ideal intellectual forms, Neoplatonism added a spiritual dimension, proposing that it is the duty of the soul to rise above its sensory environment in order to aspire to a vision of higher things and the divine "One" from which all things derive. This philosophical concept would go on to influence Christianity and lead to the theological conviction that carnality inhibits our spiritual salvation, a salvation that can only be accomplished by declaring war against our fleshly appetites. Gnosticism, which originated among the Jews of Hellenistic Egypt and spread to Egypt's early Christian community, held the view that there is a special kind of spiritual knowledge that can be obtained only by personal experience, a transcendent knowledge that is the only true path to salvation.

This concept of a secret, transformative knowledge known only to the initiated, a knowledge that could change what was base into what was pure, would radically alter chemistry's direction. It changed from a scientific to a mystical pursuit in which a coterie of practitioners, following arcane formulas, attempted to transform base metals into precious ones. Indeed, some of these formulas, or chemical "recipes," still survive, preserved in ancient Egyptian papyri dating to the early Roman Empire (the Stockholm Papyrus and the Leyden Papyrus) as well as in the fragmentary "lab notes" of such Greco-Roman alchemists as "Pseudo"-Democritus and Zosimus.

Linguistically, the art called "alchemy" echoes the ancient name of Egypt (*kemet*, "the black land"), the land of dark, fertile soil in which the art germinated. The *al* in *alchemy* was later added by its Islamic practitioners as a prefix, for in the Arabic language *al* means "the." Stripped of that prefix, *chem* would survive as the root of the modern word *chemistry*.

GOLD FEVER

The chief goal of ancient alchemists was to create gold. To the ancient Egyptians, gold was the warming color of the sun, the radiant source of all life. The flesh of the immortal gods, moreover, was believed to be made of gold, the one substance the Egyptians knew did not decay. The masks and coffins of their mummies were therefore gilded or crafted of solid gold to signify that the souls of the departed had become immortal like the gods. By a process of cultural assimilation, even the Greco-Roman residents of Alexandrian Egypt had begun to adopt the practice of using gilded mummy cases and masks for their own funerals. And to poets both Greek and Roman, the earliest and most perfect age in humankind's history had long been celebrated as golden. What better symbol, then, of humanity's quest for immortality and perfection than gold?

Of course, for those of a more mercenary temperament, the thought of transforming baser metals into precious gold—or at least fooling people into *thinking* it was pure gold—was too tempting to resist. In fact, to stop counterfeiting, the third-century CE Roman emperor Diocletian had to order all alchemical manuals burnt. Alas, even spiritual quests can be corrupted by humanity's baser instincts!

Pre-Socratic philosophy had already laid the groundwork for alchemy by asserting that all matter was one, albeit with different proportions of earth, air, fire, and water. All one had to do, then, was alter the proportions within a given sample of matter, and its properties would change as though by magic. The trick was to "coax" one material into turning into another by gently guiding it through a series of critical steps. (The ancients, as you can tell, did not appreciate the intransigent nature of chemical elements and their persistent and distinctive atomic personalities.)

Gold posed special problems to the alchemist or forger because it was so heavy. If copper could somehow be made to look less red (by mixing it, for example, with other metals like tin or zinc), it might simulate the color of gold, but it would weigh significantly less. An alternative was to blend real gold with copper and then apply sulphides to leach out the copper on the alloy's surface. Silver, on the other hand, was easier to replicate because its weight was comparable to copper's. Accordingly, copper was blended with tin, zinc, or lead to give it a more silvery look, or it was superficially whitened with a rinse of white arsenious oxide. One of the alchemists' most celebrated feats was *diplosis*, the apparent "doubling" of a quantity of gold by blending it with both copper and silver in equal amounts so as to retain its essentially golden color. Centuries earlier, a natural alloy of gold and silver called electrum had been used to make coins, but it was eventually abandoned for coin making because the exact amount of gold and silver in any particular sample was unpredictable.

"EUREKA!"

The art of simulating gold inspired one of the most famous stories in the history of Greek science and led to one of its most important discoveries.

According to the Roman architect and engineer Vitruvius, King Hiero II of Syracuse (270–216 BCE) decided to dedicate a golden crown to a local temple to commemorate his military victories.[3] Having awarded the contract to a goldsmith and having given him the requisite amount of gold, the king was informed afterward that the goldsmith had secretly substituted some silver for the gold before presenting the finished object d'art to his patron. After weighing the crown, however, Hiero found that it weighed exactly as much as the bullion he had initially supplied. Suspicious nonetheless, Hiero asked the Greek scientist Archimedes, who was then living in Syracuse, to test the purity of the crown, the only condition being that he not melt it down.

The solution came to Archimedes one day as he took a bath. As he stepped into the tub, the water displaced by his body splashed over the edge. In a flash of inspiration, Archimedes realized how he could solve the mystery of the crown. In his excitement, so the story goes, he burst naked out of his house, shouting, "Eureka! Eureka!" (I've found it! I've found it!) as he ran down the street (plate 3).

Back home in his workshop (and suitably clothed), Archimedes immersed a piece of solid gold into a container filled to the brim with water, and measured how much water spilled over the edge. Next, he did the same with an equal weight of silver. Because gold is more dense than silver, a piece of gold that weighs the same as a piece of silver takes up less space and thus displaces less water when submerged.

Next, to test the crown to see if it was solid gold, Archimedes submerged a piece of pure gold that weighed as much as the crown and measured how much water spilled out of the container. Then he submerged the crown itself. This time more water spilled out, proving that the crown was bigger than it should have been if gold were its only ingredient.

Archimedes reported his findings to the king, who promptly rewarded him. As for the fate of the devious goldsmith, it's doubtful he ever plied his craft again, let alone saw another day! Yet, in the process of uncovering the artisan's duplicity, Archimedes had discovered the principle of specific gravity.

Fascinated with the phenomenon of flotation, Archimedes also deduced that an object floats because the weight of the water it displaces is greater than its own weight. Stated another way, the weight of the fluid an object displaces forces the object up and keeps it afloat. Should the object be heavier than the fluid it displaces, it will sink. This principle, known as Archimedes's principle, not only explains why boats float, but also explains why helium balloons rise—because the helium that fills the balloon is lighter than the air that surrounds it.

MARIA THE JEWESS

Many alchemists were not as nefarious as our wily goldsmith, but they did sincerely believe in the mystical potentialities of their art. One of the most renowned of these alchemists was Maria the Jewess, "the founding mother of western alchemy." Zosimus, the greatest Greco-Roman authority on alchemy, referred to her reverentially as the "divine" Maria. Significantly, she is one of four female chemical engineers from the Hellenistic age whose names survive; the others are Cleopatra (not the queen), Paphnutia, and Theosebeia, Zosimus's own sister. The fact that Maria (or Miriam, her rightful Hebrew name) was Jewish (as Zosimus and Theosebeia may also have been) reinforces what Zosimus himself tells us elsewhere: that the Jews living in Hellenistic Egypt had learned the techniques of chemistry from the Egyptians, but they had transformed the science into a mystic art under the tutelage of God. Since there is no trace of alchemy in Jewish tradition prior to the Hellenistic period, we may be justified in attributing its rise to the influence of Hellenistic philosophical thought on Alexandria's Jews, coupled with the influence of the Gnostic tendencies that were then present in Judaism. Maria's own mysticism is revealed in a cryptic utterance attributed to her, in which she described the principle underlying the transmutation of matter: "One becomes two, two becomes three, and by means of the third the fourth achieves unity; thus two are but one." Known as the Axiom of Maria, the notion profoundly influenced psychologist Carl Jung, who used it in his *Psychology and Alchemy* as a metaphor to describe our need to reintegrate the scattered parts of our personalities in order to live harmonious lives.

Maria's ingenuity is demonstrated by her invention of three remarkable pieces of laboratory equipment: the *balneum Mariae* (as it came to be called in Latin), the *kerotakis*, and the *tribikos*. The *balneum Mariae* (Maria's bathtub) consisted of an outer vessel in which water was heated to boiling and an inner vessel designed to hold a substance meant to be safely heated to no higher than that temperature. In effect, what Maria invented (though some Classicists today would challenge her claim to fame) was the now-familiar double boiler commonly used for such nonmystical purposes as melting chocolate. To French chefs, it's known as a *bain-marie*. Ironically, in French parlance today, *une femme au bain-marie* is slang for a woman with a double boiler for brains, in short, a "dumb blonde"—certainly an unjustified sobriquet for Maria herself, since she was anything but dumb!

Named for the palette on which Hellenistic painters melted and blended colored wax, the *kerotakis* consisted of a flat pan inside a covered container. As the pan was heated and the substance on it melted and vaporized, the trapped vapors rose and chemically reacted with materials that were suspended above them.

The *tribikos*, another of Maria's ingenious inventions, was one of the very first stills in history. It consisted of three parts: a vessel in which a chemical mixture was heated, a closed cooling chamber in which the vapor condensed, and three tubes through which the distilled liquid poured out into a catch basin.

Maria's chief aim in using her equipment was to create gold from other metals. In a very real way, to an alchemist like Maria and to her contemporaries, truth was in the eye of the beholder: if it looked like gold and felt like gold, in the absence of all other tests, it *was* gold. After all, Archimedes would not discover the principle of specific gravity for telling gold from silver until sometime in the third century BCE, but even that discovery would not have dissuaded the dedicated from pursuing the quest of turning other metals *into* gold.

Maria seems to have done this in two ways. The first method was to heat copper (or copper and lead) over sulfur in the hope that the sulfurous vapor would make the "shadow" in the copper disappear and leave a residue of pure gold behind. Another was to make an amalgam of copper and 13 percent mercury, a method that has actually been used by jewelers in modern times to simulate real gold. Maria is also said to have succeeded in making precious stones "glow," either by superheating them or by treating them with a phosphorescent substance.

ALCHEMY'S LEGACY

Hellenistic alchemists were responsible for some eighty other types of chemical apparatuses employed in modern laboratories. They include baths, beakers, burners, crucibles, dishes, filters, flasks, furnaces, jars, ladles, mortars and pestles, pans, phials, stirring rods, and strainers. Even though these ancient prototypes were originally invented to serve purposes that would not be termed strictly scientific today, the mystical goals of the alchemist unquestionably led to practical advances that propelled the progress of chemistry as a true science.

As easy as it might be from our chronological perspective to deride the misguided efforts of the alchemists, we need to acknowledge that their underlying mission to interpret and manipulate the fundamental structure of matter still remains at the very heart of science. Indeed, alchemical research would persist for almost two thousand years after the Hellenistic age, and no less a genius than Sir Isaac Newton would engage in exploring its mysteries, conducting countless experiments, and writing more than half a million words on the subject.

In fact, the ancient goals of the alchemists are still alive, though entrusted to the hands of others. The search of later alchemists for an "elixir of life" to stave off death continues today in the efforts of genetic engineers to redesign human DNA; while the quest to transmute matter has already been fulfilled by physicists using cyclotrons, nuclear reactors, and particle accelerators, leading to the creation of at least twenty-nine synthetic elements. Significantly, it is from just such a synthetic element—plutonium 239—that the fate of today's civilized world now hangs.

Chapter 8

GEOGRAPHY AND GEOLOGY

I t is only natural that the ancient Greeks should have founded the science of geography, or *geographia* ("the graphic representation [*graphia*] of Earth [*ge*]") to give it its proper Hellenic name, for only by knowing what lay beyond themselves could they know their own place in the scheme of things. And only by mapping that spatial reality could they apply what they had learned to their lives. Never ones to mistake the part for the whole, their thinkers valued perspective above all else.

THE ANCIENT MARINERS

Surrounded by the sea and composed of a mainland and multiple islands, the land of Greece invites maritime travel, and, once equipped with sailing ships (having borrowed Egypt's invention of the sail), the Greeks became fishermen, merchants, and even raiders of other lands. In those days, navigation was done by hugging the shoreline or aiming at a prominent landmark, though celestial navigation was practiced as early as Homer's age of heroes. Before the days of the Trojan War, a Greek trading outpost was planted at Ugarit on the Syrian coast, and Mycenaean ships may well have sailed as far as Spain and the British Isles in search of tin, a mineral not native to Greece but necessary for making bronze, the principal metal then used for weapons and tools. Yet as Odysseus and his seafaring countrymen would be quick to testify, Poseidon, god of the sea, could be capricious, ever eager to blow sailors off course with unpredictable storm winds. In such circumstances, Greek mariners became accidental tourists until they got back home to tell the tale of their exploits, no doubt adding a fair measure of exaggeration, as would befit the survivors of Poseidon's wrath (fig. 22). With the fall of the Myce-

Figure 22: Odysseus (Ulysses) tied to the mast of his ship as winged Sirens tempt him. From an early fifth-century BCE vase in the British Museum. From Harry Thurston Peck, ed., Harper's Dictionary of Classical Literature and Antiquities *(New York: American Book Company, 1896).*

naean Empire, waves of Greek emigrants fled for safety to the haven of Turkey's western coast. Once the dark ages ended, new incentives for commerce arose as Phoenician merchant ships transported exotic trade goods to Greece, and Greek entrepreneurs in search of profit founded a trading colony on the Nile delta. As Greece's population began to increase, cities dispatched their citizens to plant commercial colonies in southern Italy and Sicily and on the shores of the Black Sea. Eventually, except for the rivalry of the powerful Phoenician colony of Carthage in the west, the Mediterranean became, for all intents and purposes, a Greek lake.

Curiosity and the spirit of adventure led a number of Greeks to venture to even farther shores. Their discoveries, like those of the legendary Odysseus, were initially recounted in oral form but were soon set down in writing in narratives called *periploi* (sg. *periplous*). These were itineraries of the places they had visited or seen, amplified by commentaries on the physical features of those lands, their flora and fauna, and the people (if any) who inhabited them. Though almost all of these eyewitness accounts have long since been lost (as, in fact, has most of ancient Greek literature itself), their details were remembered and preserved by later Greek and Roman geographers and historians whose works survive in whole or in part.

The gateway to this wider world of exploration was the Strait of Gibraltar, a watery passage flanked by two prominent peaks thirty miles apart known in classical times as the Pillars of Hercules. According to tradition, they had been erected by the hero Hercules (Heracles, in Greek) at the western extremity of his twelve heroic labors. Beyond lay the Atlantic Ocean, named for Atlas, the mythical Titan who lived at the edge of the world and held up the heavens on his shoulders.

The earliest documented voyage into the mysterious realm of the Atlantic was undertaken around 650 BCE by the captain of a merchant ship, a man named Colaeus, who hailed from the Aegean island of Samos. According to the fifth-century BCE historian Herodotus, Colaeus was blown off course and, after passing through the Strait of Gibraltar, made landfall on the southwestern coast of Spain, from which he returned with a cargo of precious silver weighing a ton and a half.[1]

About a century and a half later (ca. 530 BCE), another Greek ventured into the Atlantic, but this time on purpose. His name was Euthymenes, and he sailed from the now-French city of Marseilles (ancient Massalia), then a Greek colony, with orders to scout out promising trade routes. According to the second-century CE writer Aelius Aristides, Euthymenes turned left after leaving the Strait of Gibraltar, skirted the shoulder of west Africa, passed the Canary Islands, and got as far south as the Senegal River, where he reported seeing crocodiles and hippopotami sporting in its waters.[2]

A much more ambitious expedition was sent out by Marseilles' commercial rival, Carthage, in about 500 BCE. Led by no less than Carthage's own king, Hanno, the expedition is said to have included sixty ships and thirty thousand potential male and female colonists.[3] After encountering savages and passing the Senegal River, still amply stocked with hippos and crocs, the expedition witnessed rivers of fiery lava flowing out to the sea. Reaching the equator, Hanno came upon an island populated with hairy men and women whom his knowledgeable native interpreters called "gorillas." Unable to capture the "men," who mounted crags and defended themselves by throwing rocks, the explorers seized three of the "women," mercilessly killing and skinning them and retaining their hides as souvenirs. With their provisions failing and finding no suitable place to settle, the Carthaginians turned about and headed back to home port.

For almost two centuries, no further forays into the Atlantic took place until another citizen of Marseilles, a Greek named Pytheas, passed through the Strait of Gibraltar and, instead of turning left, turned right and sailed due north. The spirit of Pytheas's fantastic voyage is captured here by Duane W. Roller:

> Sometime in the 320s BC, a Greek explorer and scientist stood at the northern end of Scotland. It was midwinter, and he determined that the sun only rose four *peches* above the horizon (somewhat less than two meters), an observation that eventually yielded him the latitude. Then he headed north into the Atlantic, recording phenomena that were completely alien to anyone from the Mediterranean. Before returning home to Massalia, this traveler, Pytheas, had seen the frozen ocean and the midnight sun, and had theorized about the tides and determined the location of the celestial pole. Yet like many explorers to exotic places in both ancient and modern times, his reputation came to be derided and dismissed.[4]

Pytheas may have circumnavigated Britain (he was the first outsider to call it by that name) and he may have even sighted Iceland, an island he referred to as "Thoule." Yet because Pytheas's astounding observations were rejected as fantasy by later writers, his actual words (from his treatise titled *On the Ocean*) exist only in the form of scattered quotations, mostly in the works of his critics, such as the Greek geographer Strabo.

Another author who has long been looked upon as a weaver of fantasy is Homer, the name we give to the master poet of the *Odyssey*. But despite the giants, monsters, witches, and ghosts that inhabit Odysseus's ports of call, believers ever since Roman times have tried to identify the points on the hero's itinerary with specific locations in the Mediterranean by matching them up with spots that have similar topographical or ecological features. However, if we instead track the course Odysseus sailed, factoring in wind directions and the number of days he was at sea (using information in the *Odyssey* itself), we find that the final leg of his voyage—from Calypso's island to the land of the Phaeacians and then to Ithaca—may have stretched some 2,400 nautical miles from northwest to southeast. If that is true, and Ithaca was an island in the Mediterranean as tradition holds, Odysseus would have had to sail across the landmass of Europe to get there. A more geographically acceptable but radical scenario would make Ithaca not an island in the

Mediterranean but a Mycenaean trading post on the littoral of the Atlantic, a theory that actually accords with statements in the *Odyssey* that imply that Odysseus's kingdom was far removed from the traditional centers of Mycenaean power in Greece. If so, the voyage of Odysseus, however fantastic, may recall the very first Hellenic forays into the uncharted Atlantic, forays made a thousand years before Pytheas.

It was not only the Atlantic that Greek mariners explored but also the waters of the Middle East. According to Herodotus, around 520 BCE a captain named Scylax was commissioned by King Darius I of Persia to navigate the Indus River to its mouth and then sail across the Indian Ocean to the Red Sea, a mission that lasted two and a half years.[5]

Alexander the Great's overland trek to India in the fourth century BCE inspired a number of his officers, including Aristobulus and Megasthenes, to describe the terrain they crossed, just as his naval officers did—Nearchus recounting his voyage from southern India to the Tigris River, and Onescritus describing his sojourn in Sri Lanka.

THE TELLERS OF TALES

Other Greeks are more famous for the wide-ranging travel books they wrote. Two such works from the fifth century are especially worthy of note: Hecataeus of Miletus, who composed one of the very first systematic works of world geography, *The Tour of the Earth*, an encyclopedic work that described in imaginary fashion a trip around the known world from Egypt to India; and Herodotus, the "father of history," who, in researching the background of the recent war between the peoples of Greece and the Persian Empire, visited in person many of the lands involved (fig. 23). While Hecataeus's book exists only in fragments, most of which bear just the names of places and nothing more, Herodotus's glorious work stands intact, offering us a treasury of human insights into the nations of his world: their unique customs and traditions, their monuments and achievements, and their natural wonders. Driven by an innate Hellenic curiosity about the nature of the human animal, Herodotus always found juicy gossip about historical personages immensely hard to resist. Also in a typically Greek way, when confronted with multiple explanations for the same fascinating phenomenon or event, he showed his respect for his readers' intelligence by presenting them with all the

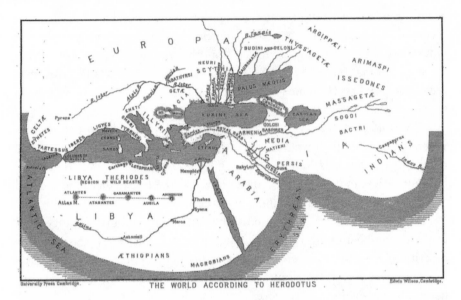

Figure 23: Map illustrating Herodotus's concept of the world. From Henry Fanshawe Tozer, A History of Ancient Geography (Cambridge: Cambridge University Press, 1897).

options, at times disclosing his own personal preferences, at other times suspending judgment entirely and leaving the matter to their own discretion.

The most extensive extant ancient work on geography, however, and our richest source of ancient geographical knowledge, is one we owe to a Greek who lived in the days of Augustus Caesar, the seventeen-volume *Geographica* authored by Strabo, who luckily had access to so many of the works that we are unfortunately denied. In addition to providing copious details about the topography of the then-known world (fig. 24), Strabo also commented amply on the flora, fauna, and ethnography of the various lands he described. In a characteristically Hellenic way, as his preface indicates, Strabo regarded geography not simply as a science but as a branch of philosophy, the Greek quest for universal wisdom:

> The science of Geography, which I now propose to investigate, is, I hold, quite as much as any other science, the business of the philosopher; and the fact that I don't take this position lightly is justified on a number of counts. First of all, the earliest men who displayed an eagerness to explore such a subject were, in fact, philosophers. . . . Secondly, wide learning, which alone

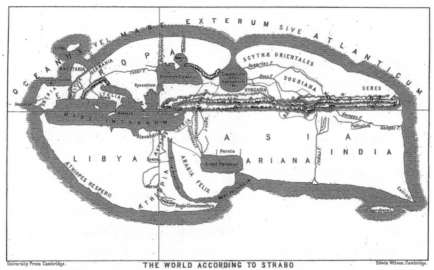

Figure 24: Map illustrating Strabo's concept of the world. From Henry Fanshawe Tozer, A History of Ancient Geography (Cambridge: Cambridge University Press, 1897).

makes it possible to undertake such research, is the province of not just any man but rather of an individual who has examined matters both human and divine—the knowledge of which, they say, constitutes philosophy's domain. And so, too, the manifold applications of geography . . . [for matters concerning war and peace and the proper use of natural resources] points to the very same kind of man, one who is devoted to reflecting on the art of life, or happiness.[6]

THE SHAPE OF THE WORLD

To judge by Homer's words, the Greeks in his day thought the world was a flat disk, bordered at its rim by a circular river called Ocean.[7] This belief persisted for centuries and was maintained by such leading sixth-century BCE figures as Anaximander and Hecataeus of Miletus. Herodotus, however, regarded the theory as ludicrous. "As for me," he said, "all I can do is laugh when I see all those folks drawing maps of the world that confound common sense, with the Ocean running around the Earth, and the Earth looking like an exact circle that somebody drew with a compass, and Europe and Asia the same size."[8]

What Herodotus hadn't fully appreciated, however, was how hypnotically appealing geometric perfection and symmetry were to his fellow Greeks. Ever since the days of Homer, symmetry had been a defining principle of literary style. Works like the *Iliad* and the *Odyssey*, for example, were endowed with a symmetrical design, an architectural plan in which analogous or contrasting elements in the story (like the arrival and departure of a character) were arranged in pairs to create concentric patterns framing a central scene.[9] In visual terms, this balanced style was also common in vase painting, one of the most important forms of Greek art. In fact, as Sir John L. Myres long ago demonstrated, even Herodotus himself used a symmetrical design when he outlined his own great history.[10] But when it came to defining the shape of the real world, the skeptical Herodotus seems to have had his doubts, perhaps because all the national idiosyncrasies and ethnic anomalies he had encountered on his global trek defied any Hellenic penchant for neatness or abstract perfection.

In actual fact, it would be this Greek fascination with abstraction that would lead to a total reevaluation of the shape of Earth, one that, as it turns out, came close to matching the real thing. Under the influence of a new philosophical movement, Earth came to be conceived of not as a flat disk but as a sphere.

It is hard to say which philosopher-scientist was the first to come up with this concept. It could have been Pythagoras, a mathematician born in the middle of the sixth century BCE, who, after migrating from Greece to southern Italy, founded a spiritual commune dedicated to the mystical study of numbers as the core of reality. Pythagoras, we are told, believed that the universe was spherical and that this spherical shape was reflected in the shape of Earth. Another candidate is Parmenides, a fifth-century BCE philosopher influenced by the Pythagorean movement. Parmenides is said to have believed that Earth was spherical because, since every one of the points on a sphere's surface is equidistant from its center, a sphere is the perfect shape. Still another possibility is the fifth-century BCE philosopher Philolaus.

Choosing the right candidate is immensely difficult because we have only fragments of their own works and therefore must rely on the testimonials of other ancient writers, some of whom lived centuries later. The key thing for us to note here is that none of these thinkers depended on empirical evidence to reach their conclusions. In the modern sense, therefore, they were not true scientists but rather based their conclusions on intellectual speculation grounded in the conviction that reality is not chaotic but rational.

This conviction was, in fact, a solipsistic projection of their own logical bent of mind: that is, they saw in external reality a humanistic reflection of themselves in the same way that earlier Greeks had envisioned the forces of the natural world as anthropomorphic gods. Yet, imperfect as it was, their vision of Earth's shape turned out to be essentially correct because their underlying conviction was, for the most part, right: an internal logic *does* pervade the appearance and behavior of the natural world, just as symmetry underlies the structure of matter in both its manifestations: organic (in the bilateral symmetry of living creatures, like the body of man himself) and inorganic (in the structure of crystals and molecules).

THE HISTORY OF THE EARTH

In the beginning, the Greeks conceived of geological history in mythic terms. As Hesiod would recount, after emerging from Chaos:

> *Gaea (the goddess of Earth) gave birth to her equal,*
> *Starry Heaven (Uranus), in order that he might cloak her*
> *And forever be the safe abode of the blessed gods.*
> *And she gave birth to the long hills, the blessed haunts*
> *Of the Nymphs who inhabit the rolling glens.*
> *And she also bore the Sea with its raging billows,*
> *Pontus, with a passion unrequited . . .*[11]

The earliest philosophers of Greece, however, dispersed the mists of mythology with the sunlight of reason, displacing the supernatural forces that had populated earlier accounts of Earth's creation with newer naturalistic explanations. As early as the late sixth century BCE, thinkers like Xenophanes and Heraclitus hypothesized from the remains of fossilized seashells found on hilltops that the surface of Earth had not always looked the way it did, reasoning that the waters of the Mediterranean had once covered the land. In the fourth century BCE, Aristotle offered a theory to explain earthquakes naturalistically instead of attributing them, as had his ancestors, to the willfulness of the god Poseidon. Meanwhile, a variety of Greek thinkers over the course of two centuries speculated on what caused the annual flooding of the Nile, with explanations ranging from heavy rains to melting snows to the waters of the

Atlantic penetrating to the heartland of Africa. Still others rationally pondered what rocks were made of, why deserts were dry, and why the tides ebb and flow. Other cultures might have been oblivious to change, or responded to it with primitive fear and awe, but for the Greeks, a world animated by dynamic change stimulated an innate curiosity that led them to a search for an explanation that would make their world more intelligible.

THE MAPMAKERS

The need to comprehend what lay on Earth's surface and to use that knowledge to traverse it led to the invention of maps. These maps drew upon both everyday experience and the expertise of ancient merchants and explorers.

The very first mapmaker whose name is preserved in Greek tradition is the pre-Socratic philosopher Anaximander (sixth century BCE), who drew Earth as a flat disc. Sharing this view of world geography was Hecataeus, who elaborated upon and annotated his fellow countryman's drawings.

Later, Pythagoras, Parmenides, and Eudoxus described Earth as spherical, dividing it into five climatic zones: a torrid zone in the middle with parallel temperate and frigid zones above and below. By the third century BCE, Eratosthenes superimposed on a representation of Earth's surface a rectangular grid composed of irregularly spaced vertical and horizontal lines. By the middle of the second century BCE, Hipparchus of Nicaea regularized the spacing of Eratosthenes's lines, creating parallels of latitude and meridians of longitude. In the second century CE, the Alexandrian Greek mathematician Claudius Ptolemy adapted these parallels and meridians to the shape of a spherical Earth by creating a cartographic projection of Earth's surface, in which the lines were curved like Earth itself. In so doing, he measured latitude from the equator (just as we still do) and made the prime meridian of longitude run through the Canary Islands in the Atlantic (just as, in a similar way, we now make it pass through Greenwich, England).

Apart from consulting landmarks and matching them to a map, travelers on land and mariners at sea could, with training, calculate the latitude they were at as long as they knew the date and could take a midday reading of the sun to determine its angle in the sky. Longitude was much harder to calculate, though Hipparchus devised an astronomical method for indicating it on maps by using the time when the same lunar eclipse was visible at a different place farther east or west.

Though the Greeks became the inventors of modern cartography, it is fascinating to note that, in using divisions of 360 degrees to denote distances, they were emulating the learned mathematicians of ancient Sumer who, three thousand years before, had counted in multiples of sixty, and had passed their numerical system on to the astrologers of Babylon.

THE SIZE OF THE EARTH

How big is Earth? the Greeks wondered. The solution was discovered in an ingenious but disarmingly simple way by Hellenistic scholar Eratosthenes (ca. 285–194 BCE), who would go on to become the director of the great Library of Alexandria.

At the very height of the hot Egyptian summer when the sun should have been directly overhead, Eratosthenes noticed that the obelisks and sundials of Alexandria cast shadows, something that ought not to have happened if the sun had been directly above them. The next year, traveling far south to Aswan (ancient Syene), he observed the opposite phenomenon: at the time of the summer solstice, the sun cast no shadows whatsoever. Back home in Alexandria, he waited for the coming of the next summer. When the calendar told him the solstice had finally arrived, he measured the angle between a *gnomon* (the pointer of a sundial) and the end of its shadow. He reasoned that the angle was the result of Earth being curved. In effect, the obelisks at Alexandria were "tilted" away from the sun by the bend in Earth's spherical surface. Since the figure he got, 7 degrees, was 7/360 of the planet's circumference, the total circumference of Earth should have been the distance between Alexandria and Aswan multiplied by the number of times 7 goes into 360 (approximately 50). Remembering that his journey by camel between the two cities had taken 50 days, and that the camel driver had told him they averaged 100 ancient "stadia" a day, the distance between the two cities should have been about 5,000 stadia, or 925,000 meters (A "stadium" was generally considered to be about 185 meters long—the distance of an ancient Greek footrace and the origin of the word *stadium*, where the races were run.) Multiplying this distance by 50, Eratosthenes got 46,250 kilometers for the polar circumference of Earth, a figure off by only 6,242 kilometers (or about 3,879 miles) from today's measurement. Given the primitive nature of the equipment and means at Eratosthenes's disposal (Alexandria is not exactly due north of Aswan, the distance between them is less than his camel driver's estimate, and the angle of the sun was a bit less than he thought), his was a truly amazing achievement.

About a century and a half later, the Stoic philosopher Posidonius (ca. 135–51 BCE) would take comparative sightings of the bright star Canopus from both Alexandria and the island of Rhodes and essentially confirm Eratosthenes's calculations.

THE HELLENIC HERITAGE

When Odysseus visited the Land of the Dead, a ghost named Tiresias told him he would have to make another journey after returning home, a journey to a distant land where men had never seen the sea. For Homer, the itinerant bard who literally sang for his supper by traveling from village to village with his bag of traditional tales, this no doubt was the setup for a "sequel" and a way of inducing his hosts to invite him to come back another time.

But for us the story of this "second" *Odyssey* serves as a metaphor for the role the explorations of the ancient Greeks have played in our lives. Were it not for their curiosity about the outer world—what Shakespeare once termed "the undiscover'd country"—the Americas might not have been found by Columbus. Surely once the Greeks conceived of Earth as a globe and recognized that its other "side" was empty, the stage was set to fill in the blanks with further adventures.

Two centuries before Columbus, his fellow countryman Dante had even imagined a senior citizen Odysseus sailing due west beyond the Pillars of Hercules in one final mission to satisfy his unquenchable curiosity about the world. For the Christian poet Dante, such a humanly inspired voyage was an act of rationalistic hubris doomed to fail. But in a wider way, it did not fail, since the driving curiosity of the Greeks coupled with the courage of medieval mariners would point the way to farther shores. Our landfalls on the moon and on Mars are but cosmic extensions of that great adventure that began some twenty-five centuries ago with the explorers and mapmakers of Hellas.

Chapter 9

METEOROLOGY

W ind and weather play a key role in two of Greek literature's most ancient works: Homer's *Odyssey* and Hesiod's *Works and Days*.

THE KEEPER OF THE WINDS

On their voyage home from the Trojan War, Odysseus and his men landed on an island ruled by a king named Aeolus. Aeolus had been appointed Keeper of the Winds by the great god Cronos. After graciously showing his visitors hospitality, Aeolus sent Odysseus and his men off with an unusual parting gift: a tightly tied leather bag filled with wind.

After leaving Aeolus's isle, the Greek warriors sailed for nine days, and on the tenth saw their homeland on the horizon. But when Odysseus, who had been manning the sails all this time, fell asleep from exhaustion, his greedy crew, suspecting the leather bag actually held treasure, made the decision to open it up.

> As they untied the bag, all the winds rushed out,
> and a hurricane seized them,
> carrying them out to sea as they wailed,
> driving them away from their homeland.[1]

This would not be the last time Odysseus's plans would be thwarted by storms at sea, for his sworn enemy was Poseidon, the god of the very element over which he had to travel if he was ever to see his family again.

A GREEK FARMER'S ALMANAC

Bad weather could prove devastating not only to Greek mariners but also to Greek farmers. Homer's compatriot Hesiod, who knew from personal experience what a peasant's life was like, incorporated his knowledge into a poetic farmer's almanac called *Works and Days*. Of all the seasons, he tells us, winter was the worst. His advice:

> *Shun harsh January and February*
> *and their cruel frosts*
> *when Boreas, the North Wind,*
> *blasts the earth,*
> *roiling the broad sea,*
> *and making the woods howl,*
> *toppling the lofty-leafed oak*
> *and the bristling pine.*
> *He even penetrates an ox's hide,*
> *sparing only thick-fleeced sheep,*
> *and turns an old man into tumbleweed.*[2]

METEOROLOGY AND RELIGION

From practical necessity, Greek farmers and sailors had to be weather-wise. But as the names Aeolus, Poseidon, and Boreas show us, they saw the hand of the divine in the weather they faced.

Even Zeus, the king of the Olympian gods, was a meteorologist of sorts (fig. 25). Ensconced as a sky-god in his palace atop Greece's highest mountain, he was called the Cloud-Gatherer and He Who Delights in Thunder, and he wielded the lightning bolt as his weapon against evildoers. In time of drought, the pious prayed to him for rain.

Given the spiritual implications that weather had for the Greeks, when did they first begin to seek a rationalistic rather than a religious explanation for it?

Though the pre-Socratic philosophers discussed some aspects of Earth's atmosphere, the earliest-surviving books that deal exclusively with weather

Figure 25: Portrait of the sky-god Zeus. From Estelle M. Hurll, Greek Sculpture: A Collection of Sixteen Pictures of Greek Marbles with Introduction and Interpretation *(Boston: Houghton Mifflin, 1901).*

date to the middle of the fourth century BCE, some four centuries after the time of Homer and Hesiod. The authors are Aristotle (384–322 BCE) and Theophrastus (371?–287 BCE).

Why did it take Greek thinkers so long to focus on this subject? The explanation may be traced to the awesome power of atmospheric phenomena, the Greeks' deep connection with Zeus, and the dependency of an agrarian population upon the blessings of the gods. After all, as late as 399 BCE the philosopher Socrates had been convicted of and executed for having rejected the traditional gods of Athens and for having corrupted young Athenian men into following his atheistic views by using a seductive form of rationalism. Perhaps, then, no field of science was more fraught with danger for Athenian intellectuals than the seemingly (to us) benign field of meteorology, "the study of what is high in the air" (from Greek *meta*, "above" and *aeiro*, "lift"), since it trespassed on a province traditionally governed by Greece's supreme god.

PHILOSOPHERS AS WEATHERMEN

For Aristotle, as for other Greek thinkers, meteorology included more than weather, embracing as it did not only the topics of temperature and humidity; wind and clouds; rain, snow, and hail; and thunder and lightning, but also such "high in the air" phenomena as rainbows, comets, shooting stars, and even the Milky Way. While Aristotle's efforts to explain these phenomena were logical, his theories were essentially persuasive inductive arguments based principally on pure logic rather than on empirical evidence or experimentation. The one important assertion he makes that is scientifically valid is his explanation for the cyclical nature of rain: that the heat of the sun causes moisture on Earth to evaporate; that the vapor then rises and, as it cools, condenses into clouds; and that the condensed liquid then falls to Earth again as rain.

Following in the footsteps of his friend and mentor Aristotle, Theophrastus continued the rational inquiry into atmospheric phenomena. In his book *On Weather Signs*, he discussed how to forecast weather using signs evident in nature, such as the coloration of the moon or the behavior of certain animals. His information, he makes clear, was based both on his own personal observations and on the experience of Greek farmers and herdsmen who lived close to nature. For predicting local weather, he counsels, we should be sure to heed the signs relied on by local residents. While most of the correlations he presented sound more like folklore than science ("It is a sign of coming rain when frogs are more vocal than usual."), Theophrastus nevertheless tried to apply a specificity and deductive rigor to his research, which was largely absent from Aristotle's *Meteorology*. His down-to-earth advice about weather was even given a poetic touch in the third century BCE by Aratus, who concluded his poem *The Phenomena* with stanzas describing how nature can warn us about bad weather—if, for example, we hear mice squeaking louder than usual or see dogs digging an inordinate number of holes in the yard. Indeed, it may turn out that a few "signs" modern city dwellers would be prone to scoff at merit further scientific investigation by naturalists, reflecting as they do ancient but time-tested folk wisdom.

Building upon a pre-Socratic notion that wind is simply air in motion (as opposed to a confluence of exhalations from Earth, as Aristotle had believed), Theophrastus went on (in *On Winds*) to advance new theories to describe the movement of air currents, including the role of the sun's heat in setting air into motion and the effects of topography on wind. He also attempted to explain climate change and local climatic differences, combining his own distinctively deductive method with keen powers of observation.

THE TOWER OF THE WINDS

The greatest monument to ancient Greek meteorological research is a forty-foot-tall octagonal structure made of marble that still stands in the shadow of Athens' Acropolis (fig. 26). Designed by the Hellenistic astronomer Andronicus of Cyrrhus in the first century BCE (though it may in fact date back a century earlier), the so-called Tower of the Winds gets its name from the sculpted personifications of eight male wind-deities (including the North

Figure 26: Athens' Tower of the Winds as it may have looked in antiquity. From Christopher Wordsworth, Greece: Pictorial, Descriptive, and Historical *(London: Orr, 1840).*

Wind, Boreas, and the West Wind, Zephyrus) portrayed in a belt-like frieze around the tower's top (fig. 27). Surmounting the building's conical roof, a weather vane once spun. Each of the monument's eight exterior sides is still inscribed with the lines of sundials, while inside the structure stand the surviving stone fragments of an elaborate water clock (or *clepsydra*). Thus the Tower of the Winds was both an ancient timekeeping station and a meteorological research center all in one.

Figure 27: Sculptural personifications of the North Wind (above, holding a conch shell for blowing) and the South Wind (below, holding a rain-filled pitcher) from the Tower of the Winds. From Franz von Reber, History of Ancient Art, *rev. ed., trans. and augm. Joseph Thacher Clarke (New York: Harper & Brothers, 1882).*

As old Boreas, wrapped in a wool cloak, blows a chill blast from his conch shell and young Zephyrus floats through the warm air bearing springtime blossoms, the Tower of the Winds signifies both the mythic opening act and the scientific finale of meteorology in ancient Greece.

Chapter 10

ASTRONOMY

O f all the scientific subjects the ancient Greeks studied, the earliest may have been astronomy, for their observations are recorded in their oldest-surviving literary texts, poems that date back almost three thousand years.

EPIC SKIES

Homer's *Iliad*, the story of the Trojan War, tells the tragic story of Greece's greatest warrior, Achilles. The epic also contains one of the most elaborately detailed poetic descriptions of an ancient warrior's armor, the shield of Achilles, handmade by none other than Mount Olympus's divine metalsmith, Hephaestus.

Using bronze, tin, silver, and gold, the god portrayed the world Achilles knew intimately, for in carrying that fateful shield into battle Achilles would be surrendering his very life. Hephaestus depicted cities at war and at peace, fertile fields and pasturing herds, and dancing maidens and men but also the universe that surrounded the activities of the human race:

> On it he wrought the earth, and the sky and the sea,
> the tireless sun and the full moon,
> and all the constellations with which the sky is crowned:
> the Pleiades and the Hyades and mighty Orion,
> and the Bear they also call the Wagon
> that turns round there keeping an eye on Orion
> and never dips into the Ocean.[1]

As noted earlier, the companion poem to the *Iliad*, Homer's *Odyssey*, describes the aftermath of the Trojan War and the troubled homeward voyage of

Odysseus. Shipwrecked on the island of the seductive goddess Calypso, Odysseus longed for the day of his return to his beloved family and kingdom. Finally, Zeus took pity on the hero and commanded Calypso to release him. After giving him fresh supplies and tools to build a new boat, she sent him on his way:

> Joyfully the hero Odysseus spread his sails into the wind.
> Taking his seat, he skillfully manned the rudder,
> nor let his eyes surrender to sleep
> as he watched the Pleiades and the late-setting Ploughman
> and the Bear they also call the Wagon,
> that turns round there keeping an eye on Orion
> and never dips into the Ocean.
> For this constellation the lovely goddess Calypso bade him
> always keep on his left as he sailed the sea.
> For seventeen days he sailed the sea,
> and on the eighteenth day there appeared close at hand
> the shadowy mountains of the land of the Phaeacians
> rising from the misty deep like a shield.[2]

Documented here are not simply poetic observations about heavenly bodies but practical examples of their utility to a seafaring nation. For a civilization dotted with islands and surrounded by the sea, transportational necessity became the mother of invention and gave birth to celestial navigation.

The ancient Greeks, however, were not only a race of mariners but of farmers as well, farmers for whom a precise knowledge of seasonal change was critical for successful planting and harvesting. Thus the Greeks looked to the heavens as a celestial "farmer's almanac."

A successor of Homer named Hesiod (ca. 700 BCE) incorporated this agrarian information into a poem called *Works and Days*. Dedicated to a lazy and litigious younger brother named Perses, Hesiod's poem is part moralizing tract and part agricultural handbook. The public recitation of his verses was meant to help his compatriots prosper both ethically and economically. Here is a sample:

> When mighty Orion first appears, order your slaves
> to thresh Demeter's holy grain
> on a smooth threshing-floor in an airy place

and store it in carefully-filled jars. . . .
But when Orion and Sirius are halfway across the sky
and rose-fingered Dawn spies Arcturus,
that's the time to pick all the grapes,
Perses, and haul the clusters home.
Leave them out in the sun for ten days and nights,
then cover them up for five, and on the sixth day
draw off into jars the bounty of cheerful Bacchus.
But when the Pleiades and Hyades and mighty Orion
begin to set, that's the time to set your thoughts on ploughing.[3]

Hesiod's practical mindset didn't seem to have prevented him from composing another more imaginative poem, *Astronomy*, in which he recounted the myths associated with the major stars and constellations and the legendary figures for whom they were named, along with suggestions about how the heavens might influence our lives. Only fragments of the poem, however, survive.

A FELLOWSHIP OF STARGAZERS

As ancient astronomers, the Greeks were not alone, for other Mediterranean cultures were filled with wonder and awe as they beheld the myriad clusters of stars that gleamed overhead in the pitch-black darkness of a sky unspoiled by the electric lights of today's cities. No wonder, then, that the ancient Egyptians and Babylonians scanned the heavens, connected stellar dots into constellations, and wove their astrological meaning and mystery into their myths, rituals, and religions. The Egyptians, who invented the world's first solar calendar, believed that the souls of the dead revolved around the North Star, and they predicted the rising of the Nile's nourishing waters from the annual debut of the star Sirius in the east. The Babylonians, for their part, kept meticulous records of celestial events, devised the signs of the zodiac familiar to us, and raised the predictive art of astrology to new heights. In these ancient pre-telescope times, however, only five planets were visible to the naked eye.

Where the Greeks differed from the cultures of the Near East was in their own rational approach to the sky's wonders. We have already seen glimmerings of that approach in the Hellenic application of the stars to the practical needs of navigation and agriculture. But the Greek imagination, stimulated by heavenly riddles, went much further by seeking a rational explanation for the workings of the cosmos.

In large part, their quest was inspired by the fact that they viewed the universe not as something inherently sacrosanct but as a humanistic projection of themselves. If reason underlay the works of man, they argued, so must it underlay nature itself. Appropriately, the ancient Greek word for the universe, *kosmos*, meant a lovely arrangement whose beauty reflects order. Accordingly, reason became the key to unlock the universe's secrets. Because they viewed the stars and humanity as a logical continuum, ancient Greek scientists did not exclude the possibility that the former might even influence the latter. But the chief question they asked was: How is our universe designed? As it turned out, solving that structural puzzle would become their enduring preoccupation.

THE REBELS OF IONIA

The search for a solution began in a very unlikely place—not mainland Greece or its nearby islands but the far-off shores of southwestern Turkey. It was there that Greeks had fled for refuge in the eleventh century BCE, when the fall of Troy was followed by the collapse of Mycenaean civilization. And it was there in that more secure environment, around 800 BCE, that Homer himself may have been born, the poet who looked back across the post-Mycenaean dark ages and tried to poetically recapture the heroic spirit of a lost world by weaving its traditions into a grand tapestry of inspiring adventure.

The enterprising Greek colonists who settled on the Ionian coast eventually built up flourishing cities, most notably Miletus, that prospered on trade. To accelerate commerce, the Greeks capitalized on a new Lydian invention, coinage, replacing an older, slower barter economy with a faster, newer one based on money. The influx of wealth from accelerated transactions soon led to the rise of a prosperous middle class that challenged the political status quo with revolutionary populist ideas. Winds of change were now sweeping over Ionia, far-reaching winds that affected not only the way politics worked but the way Greeks throughout the eastern Mediterranean saw themselves and expressed themselves. For the first time, statues portrayed self-reliant human beings rather than gods, and new forms of poetry were invented to lyrically encapsulate the vibrant personal feelings of the living instead of the deeds of dead heroes from a bygone era. A brave new world opened up, a world made by human beings with their own hands, a world in which the Olympians took a back seat.

The radicalism of this Greek renaissance is echoed in verses from the philosopher-poet Xenophanes (ca. 570–480 BCE), who mocked humanity's penchant to bow down before anthropomorphic gods:

> *If horses and cattle had hands*
> *and wanted to make icons like man's,*
> *the horses would sculpt gods that galloped*
> *and the cattle would sculpt gods that mooed.*[4]

In such an intellectual climate of skepticism, people boldly looked at the world itself with fresh eyes, peeling away its conventional mythological veneer to reveal the underlying natural—not theological—causes of its behavior. As a result, the horse-drawn chariots of the sun and moon, driven across the sky by the solar god Helios and the lunar goddess Selene, would forever be mothballed as mere poetic figures of speech.

The ancient Greeks unanimously agreed that their first true astronomer was Thales of Miletus (ca. 640–546 BCE), the founder of natural science and the first thinker ever to rationally investigate the physical principles behind the operation of the universe. The most popular story told about Thales was recounted by Plato: Thales was so busy gazing at the heavens one night he fell into a well—proof that woolly-headed philosophers and astronomers need to keep their feet on the ground.[5] But Thales was nobody's fool. Having somehow deduced from his meteorological observations in wintertime that a bumper crop of olives was coming, he secured an option to all the local olive presses, thereby cornering the market and making a fortune in rentals when the harvest came in.[6]

In Egypt, he used his powers of deduction to calculate the height of a pyramid from the length of the shadow it cast. Thales is also credited with having been the first man to predict a solar eclipse, perhaps by identifying a repeated pattern in the intervals between eclipses of the moon and sun as recorded in Babylonian annals.[7] The eclipse in question took place on May 28, 585 BCE, and its eerie darkness so terrified the armies of the Medes and Persians, who were engaged in battle at the time, that they immediately stopped fighting and signed a peace treaty. But to Thales, the scientist, the eclipse was not an omen from the gods but simply the impersonal and predictable consequence of celestial machinery at work. By precise observation, Thales was also able to accurately predict the time of the summer and winter solstice. For all these discoveries and others, Thales was rightly ranked among the Seven Sages of Ancient Greece.

Anaximander (ca. 610–547 BCE) was a fellow Milesian and Thales's disciple. We already encountered Anaximander's name when we learned he was the first to draw a map of the inhabited world, a world that he mistakenly envisioned as a flat disc. This shortcoming, however, is more than made up for by the fact that he was the first to conceive the notion of outer space, the idea that the universe is limitless and unbounded, a concept that set his thinking apart from the Near Eastern belief that there is a heavenly vault hanging over Earth. His was a view visually confirmed only in 1969, more than 2,500 years after his death, when American astronauts looked back from the moon and saw the bright blue marble that is our home.

Anaximander also proposed that other heavenly bodies moved in circular orbits around Earth as it floats in space. Though we now know that this "geocentric" theory of the solar system and universe is incorrect, it is important to recognize how very reasonable it actually was, given the apparent rising and setting, and thus revolution, of heavenly bodies around our planet from the perspective of a person standing on Earth. It was in fact the standard view that Egyptian and Mesopotamian priest-astronomers had long accepted.

However, rather than thinking of the sun, moon, and planets as orbs, Anaximander imagined that what we see are really glowing holes in a succession of great celestial wheels that forever revolve in space around Earth. To a Greek like Anaximander, these "wheels" had to be circular because the circle was considered a perfect shape, and, because he believed in consistency and the fundamental power of prime numbers, he argued that each wheel had to be separated from one another by a factor of three. By thinking this, Anaximander was subconsciously projecting the Hellenic penchant for mathematical perfection on the universe instead of arriving at cosmic conclusions based on empirical data. In this case, at least, cultural subjectivism trumped objective science.

A strong objection, however, was raised to Anaximander's geocentric theory. This objection came not from the eastern Mediterranean but from the west, where in the sixth century BCE a mysterious figure named Pythagoras founded a colony in southern Italy dedicated not to commercial expansion but to the mystical power of numbers and the search for spiritual purity. He and his followers, most notably the later philosopher Philolaus, believed that Earth was a perfect sphere that together with the other spherical planets, an unseen "counter-Earth," and the sun itself revolved around an invisible central fire. The Pythagoreans also believed that the orbits of all heavenly bodies were separated not by a factor of three (as Anaximander had held) but by mathematical intervals that are the same as those that produce harmony among the strings of a musical instrument. Such cosmic harmony, they asserted, generated in space a "music of the spheres" inaudible from Earth. Here, as earlier, we encounter an astronomical theory based not on evidence from the universe itself but superimposed on it by a human mind enamored of mathematical perfection.

Following the deaths of Thales, Anaximander, and Pythagoras, the interest of the Greeks in theorizing about the cosmos for the most part waned. Invaded by the Persian Empire at the beginning of the fifth century BCE, mainland Greeks responded by banding together and driving out the foreign aggressor. Inspired afterward by the nationalistic victory in which they had played a leading role, the Athenians entered a golden age of creativity in literature and the arts. The focus of Greek thinkers accordingly shifted from the abstract contemplation of the universe to an inquiry into human nature and humanity's capacity for glorious achievement on Earth.

One notable chronological exception who kept his eyes on the heavens was the philosopher Anaxagoras (ca. 500–428 BCE). Among other things, Anaxagoras rationalistically proposed that the sun was, in actuality, a gigantic red-hot stone, that the moon's topography was like Earth's, and that eclipses and the phases of the moon were not manifestations of the divine but were caused naturalistically by the movement of celestial bodies.

With Athens' political disintegration at the end of the fifth century BCE, philosophers like Plato (ca. 429–327 BCE) and his student Aristotle (384–322 BCE) once again turned their intellectual attention to the natural world. Because it seemed so humanistically self-evident, they accepted Anaximander's geocentric model of the universe. Plato, for his part, even saw in it the handiwork of a creator god[8]—a thought that would have made old Xenophanes roll over in his grave! Moving beyond the original four elements of pre-Socratic thought, Aristotle argued that heavenly bodies were unique in that they themselves were composed of a "fifth" element, or *quint*essence (to use its Latin equivalent, the source of our word *quintessential*).

Significantly, while acknowledging astronomy's practical utility, Plato the idealist disparaged its ultimate value since astronomy, he claimed, directs our minds only to the stars but not beyond them to higher spiritual truths that matter more.[9]

During the following century, Aristarchus of Samos (310–250 BCE) became the first astronomer to calculate the size of a heavenly body. Using principles of geometry, he estimated the sizes of the sun and moon relative to that of Earth and their distances from our planet. In addition, he took a bold and seemingly antihumanistic step by advancing a heliocentric ("sun-centered") theory of the universe. But because Aristarchus did not put humankind and his own planet at the center of things, his heliocentric theory received a generally cool reception. Indeed, we know of only one ancient astronomer, Seleucus of Seleucia, who vehemently took his side. The battle between the two sides would not be won until 1543 CE, with the posthumous publication of Nicolas Copernicus's epoch-making studies. Until then, the view that Earth was the center of the universe carried the day.

Of course, there were anomalies in the geocentric model, the most glaring being the fact that some planets appeared to exhibit retrograde motion—at times stopping and reversing their courses and then continuing on their former orbital paths. Nevertheless, various ancient astronomers such as Eudoxus of Cnidus (fourth to third century BCE) and Apollonius of Perge (ca. 200 BCE) strove to resolve such discrepancies by performing virtuoso feats of mathematical ingenuity and mechanical modeling to make the answers come out "right." The devices they employed included a set of concentric cosmic spheres rotating on different axes; planets whose own smaller orbits circled Earth in epicyclic motion; and planets that circled Earth in eccentric orbits. For his part, the third-century BCE genius Archimedes is even said to have constructed a celestial globe and a planetarium that mechanically reproduced the simultaneous movement of the sun, moon, and planets around Earth.[10]

Eudoxus's research into the location of constellations was given a boost by a fourth-century BCE poet named Aratus, who composed an elaborately detailed poem, *The Phenomena*, inspired by one of Eudoxus's astronomical treatises. So popular was Aratus's star-studded composition, "it became the most widely read poem, after the *Iliad* and *Odyssey*, in the ancient world, and was one of the very few Greek poems translated into Arabic."[11]

Unlike Eudoxus and Apollonius, other Hellenistic astronomers like Hipparchus (fl. 200 BCE) avoided going through mental gymnastics to solve problems in celestial mechanics. Instead, they focused principally on the stars themselves. Drawing on the vast library of Babylonian observations at his disposal, Hipparchus learned how to calculate where heavenly bodies would be located in the sky at any given time, and he compiled a catalog of 850 stars and their precise celestial coordinates. His painstaking observations of the heavens combined with his research into old astronomical records led him to discover the precession of equinoxes—the gradual shifting in the apparent location of the stars due to a wobble in Earth's axis. And having invented trigonometry, he used it to calculate the distance from Earth to the moon.

Furthermore, in an age before the telescope, Hipparchus used his ingenuity to invent a number of new instruments that dramatically advanced the scientific exploration of the skies.

Among these inventions were the *dioptra* and, possibly, the astrolabe. The dioptra was a kind of optical slide rule for sighting purposes that enabled the viewer to judge the relative size and distance of a heavenly body. The astrolabe was an intricate device with multiple mechanical disks, some movable and some fixed, that allowed its user to predict the relative positions of heavenly bodies and determine the time of day or night from their location.

As the distinguished French astronomer Guillaume Bigourdan wrote in Hipparchus's praise: "With this extraordinary man there suddenly appears a perfected astronomy, far superior to that of the preceding age; the theories of the sun and moon are formulated, and those of the planets outlined; the great desideratum of ancient astronomy, the prediction of eclipses, is now a problem solved. For the first time, the positions of a great number of stars scattered in the sky were known, and by the discovery of the precession their co-ordinates for any period could be calculated."[12] Hipparchus has therefore rightly been called the greatest astronomer of antiquity.

Hipparchus has a unique connection with mythology. According to Greek myth, Zeus, the king of the gods, punished Atlas for siding with the Titans in their war against the gods of Olympus. As his penalty, Atlas was forced to hold up the heavens for all eternity. For some seventy-five years, tourists visiting New York's Rockefeller Center have marveled at Atlas as he stands on Fifth Avenue, carrying the burden of the universe on his powerful metallic arms and shoulders. This modern statue, however, has an ancient precedent, originally displayed in Rome's Farnese Palace and today preserved in Naples' National Museum. Carved out of marble in the days of Imperial Rome and based on a Hellenistic original centuries older, it shows a kneeling Atlas bending under the weight of a giant marble globe on which are depicted forty-one heavenly constellations represented by their pictorial symbols—a lion, for example, for the constellation Leo and a crab for Cancer.

Though the statue with its globe has been on display for centuries, only recently has its full meaning been understood. While touring Italy, Bradley E. Schaefer, a professor of physics at Louisiana State University, paid a call on the "Farnese Atlas" and was amazed by the scientific accuracy of the globe's detailed rendering of the nighttime sky (plate 4). Not only were stars and constellations represented, but so, too, were the equator, the Tropics of Capricorn and Cancer, and the Arctic and Antarctic circles—information that, together with the relative positions of the constellations, enabled him to date when this picture of the night sky had been "snapped."

Using seventy key stars as his guides, Dr. Schaefer was able to deduce that the Mediterranean sky would have looked that way around 125 BCE, the very time the legendary Hipparchus did his research and published his star catalog. As Schaefer argued, this in turn could have inspired an anonymous Hellenistic sculptor to produce this mythological version in marble—an illustrated "Atlas" based upon Hipparchus's book.[13]

THE WORLD'S OLDEST COMPUTER

Sadly, the inventor of the most fascinating astronomical instrument of ancient times also remains anonymous. The instrument is called the Antikythera mechanism, named for the Aegean island in whose waters it was found by sponge divers in 1900, during what turned out to be underwater archaeology's earliest adventure. A calcified lump of corroded bronze when it was first uncovered, the object was originally thought to be a broken piece of a statue that had been transported across the sea in an ancient ship that had foundered and sunk in a storm. As it dried on a shelf in Athens' National Museum, the lump of bronze split open, revealing a curious multi-geared clockwork mechanism with dials marked off in degrees and inscribed with astronomical symbols in Greek (plate 6). Examined in the 1950s by English physicist and mathematician Derek J. de Solla Price, it was discovered to be a scientific instrument designed to predict the rising and setting of major stars and constellations, the phases of the moon, and the movements of the planets.

A testament to the stunning ingenuity of the Greek mind, the Antikythera mechanism is the most complex ancient scientific instrument we possess and the oldest-surviving "computer" in the world. (By computer, I mean a highly sophisticated calculating device that can automatically process numerical information and reveal correlations between interconnected sets of data.) In fact, X-ray studies undertaken decades later by Michael T. Wright of London's Science Museum revealed a far greater complexity of internal parts than had previously been imagined. In 2006 and 2008, employing advanced surface imaging and high-resolution X-ray tomography, a team of British, Greek, and American researchers led by Dr. Tony Freeth announced that they had discovered twice the number of decipherable inscriptions on the mechanism's surviving fragments and had gained new insight into the working of its internal gears.

When deciphered, some of the inscriptions suggested that the device had been used in concert with a mechanical planetarium. One set of gears measured the progress of the nineteen-year "Metonic" cycle that brought solar and lunar years back into harmony, while another simultaneously measured the progress of the eighteen-year "Saros" cycle that marked and predicted the recurrence of solar and lunar eclipses based on Babylonian data. In reckoning lunar eclipses, the gearing took into account the moon's elliptical orbit, an idea Hipparchus had first proposed. Still another set of gears functioned as a sports calendar, indicating when the Olympic Games and the other quadrennial athletic festivals of the Greeks would occur.

From the frozen coordinates of its month and zodiac slip-rings, we know the Antikythera mechanism was last set and used in 80 BCE. The names of the months in the inscriptions match the spellings traditionally used in northwestern Greece and Sicily, suggesting that Archimedes, who had lived in Sicily a century and a half or so earlier, may have had a hand in its original conception and design. Yet, exactly who the builder of the computer actually was, we will never know, nor will we know if he drowned in the sea that claimed his greatest masterpiece.[14]

THE ENCYCLOPEDIA OF THE COSMOS

The last of the great ancient Greek astronomers lived in Alexandria, Egypt, during the second century of the Roman era. His name was Claudius Ptolemaeus, or Ptolemy, for short. Besides adding another 178 stars to Hipparchus's catalog (thereby expanding the grand total to 1,028, the longest such celestial list from antiquity), Ptolemy left us the ancient world's most comprehensive and detailed exposition of ancient astronomy's methods and achievements. Originally called the *Mathematike Syntaxis* (Systemization of Mathematics), the thirteen-volume work eventually received an honorific title from the Arab astronomers who came to admire it in later centuries. They called it the "Almagest," from the Arabic word for *the* (*al*) and the Greek word for *greatest* (*megiste*), the title by which it is still known today. In the preface, Ptolemy rebutted Plato's claim that astronomy was a mere intellectual distraction, arguing instead that astronomy "makes its students into lovers of divine beauty, instilling that same quality in their souls."[15] Thus, claimed Ptolemy, humankind's rational search for knowledge, expressed through astronomy, fulfills their spiritual quest. Ptolemy's conviction that the heavens are a manifestation of the divine also led him to celebrate astrology's ability to guide our lives, a belief shared by later giants of science who read his works, including Johannes Kepler and Sir Isaac Newton. Ptolemy's most influential belief, however, was that Earth, not the sun, was the center of the solar system, a belief that would govern, and constrain, the thinking of astronomers for well over a millennium thereafter.

Despite this, Ptolemy's deepest conviction—that astronomy is no mere "science" but a portal to something higher—merits our continuing and serious reflection. As British Classicist G. E. R. Lloyd has passionately argued:

> Even among those ancient writers who are aware of the possibility of applying science to practical ends, the chief motive for the investigation of nature generally remains the non-practical one. The ancients explored nature not to dominate her, certainly not to exploit her, but to become wise. The aim was understanding: and the view was often expressed that without understanding and knowledge, peace of mind and happiness are unattainable. . . . Science was, indeed, a part of philosophy . . . an aid to improving the character. . . . We cannot, I think, fail to be struck by the ideal of science as an inquiry to be pursued as an end in itself, as part of the good life, and indeed of moral education. Greek cosmology and science are anthropocentric. But the other side of that coin is that they are also confidently humanistic, confident that the primary justification of the inquiry is not practical applicability, but knowledge. Science should be useful, but the criterion of usefulness is not material welfare, but understanding.[16]

In this pragmatic and materialistic age of ours, when the test of any career's worth is too often how much money you can make, the ancient profession of astronomy has much to teach us about what our proper quest should be.

PART II:

EXPLORING THE UNIVERSE

Section 2: The World Within

Chapter 11

BIOLOGY

Few people ever possessed as great a passion for life as did the ancient Greeks. When Achilles was honored by all the other ghosts of Hades for his extraordinary valor, the dead hero declared he would "rather be the humblest slave of the humblest peasant on earth than sovereign among the dead," just so he could breathe the breath of life again beneath a warming sun.[1] Given their lust for life and the driving curiosity that explains all their other sciences, it is only fitting that the Greeks would excel at biology, the study (*logos*) of life (*bios*), pursuing it with the same fervor their historians and dramatists showed in investigating the nature of humankind.

OBSTACLES TO RESEARCH

The scientific examination of the human body was inhibited by Greek religion, which viewed the dissection of corpses as an act of desecration. As a result, during the classical age, when so much energy was expended trying to describe the human psyche in literature and to delineate man's outward appearance in the visual arts, little attention was paid to exploring the inner workings of the human body.

Ironically, the humanistic celebration of man among the Greeks also had the effect of inhibiting the study of other "lesser" forms of organic life. This humanistic bias is evident as far back as the days of Homer and Hesiod. Homer rarely described animals and, when he did, he singled them out mainly for the traits that made them seem human—as in the case of Achilles's compassionate horses[2] or Odysseus's faithful dog.[3] The flora and fauna that appear in Homeric poetry are mostly relegated to the role of stage scenery or function in similes as minor players in a drama that features men and women as its obvious stars.

In the *Theogony* and *Works and Days*, Homer's fellow poet Hesiod became the first Greek writer to chronicle the beginnings of the universe. But, contrary to the detailed account found in the biblical book of Genesis, where the creation of other living things[4] was the prelude to God's creation of man and woman,[5] Hesiod the humanist made the creation of humankind (fig. 28) an unprecedented act by the Olympian gods.[6] A related Greek myth, preserved by Apollodorus[7] and later recounted by the Roman poet Ovid,[8] told how the sole survivors of the Great Flood—Deucalion and Pyrrha—had repopulated Earth by casting stones over their shoulders; the stones thrown by Deucalion instantly became men, and the stones thrown by Pyrrha became women. As to the repopulation of Earth with other creatures, no mention was made of them—unlike the biblical account of the Flood in which God ordered Noah to take paired creatures aboard his ark.[9]

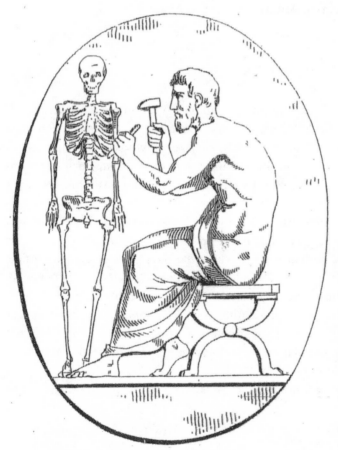

Figure 28: An ancient carved gemstone depicting the god Prometheus sculpting the prototype of man. From George Redford, A Manual of Sculpture (London: Sampson Low, Marston, Searle, and Rivington, 1882).

During the golden age in Athens this deliberate emphasis on humanity to the disregard of the rest of the natural world continued in literature and art. Apart from a simple architectural backdrop, the stage in Greek drama was bare, and landscape was absent from the painted decoration of Greek vases. The one extant exception in Greek art is the victory parade on the Parthenon frieze, which includes horses whose bones and muscles reveal a thorough understanding of equine physiology. The emphasis in classical literature was likewise on human beings, whose behavior and accomplishments were the central concern of authors. The only exception was the occasional attention Herodotus showed in his *History* to the exotic flora and fauna of foreign lands. Even so, like the Parthenon itself, Herodotus's *History* was explicitly created to celebrate the glorious deeds of man.[10]

THE BODY BEAUTIFUL

During the mid-fifth century BCE when Athens reached the apex of its political power, affluence, and self-confidence, the ideal human physique became the preoccupation of the city's greatest sculptors: Myron, Polycleitus, and Phidias. Because of the community's conservative moral sensibilities, however, only the male body was depicted naked, while the female was clothed. The chief aim of Athenian sculptors was to produce in marble or bronze a lifelike image of the ideal man, posed standing or transfixed in action. In their passion for perfection, the artists seem to have measured the bodies of a variety of male models and then averaged these measurements in an attempt to discover a universal set of anatomical proportions that would reflect the innate "blueprint" of the archetypal male. This perfect body was to be defined, however, not by the absolute size of any one particular anatomical feature (the height of the body, for example, or the size of the chest) but rather by the relative sizes of the body's parts when compared with each other. The result was a set of proportions taken from real bodies that was then refined, through a process of intellectualization, to reflect a set of artificially exact mathematical ratios.

Polycleitus, for example, wrote a book called *The Canon*, in which he described the ratios of his sculptural template and then made statues such as the *Spear-Bearer*, or *Doryphoros*, to illustrate them (fig. 29). Though the book

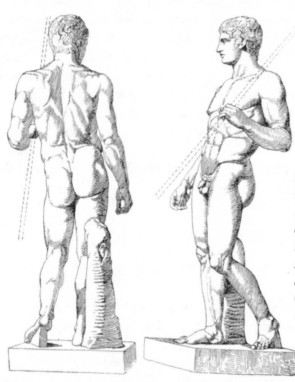

Figure 29: Marble copy of the Spear-Bearer *(Doryphoros) of Polycleitus. From Franz von Reber,* History of Ancient Art, *rev. ed., trans. and augm. Joseph Thacher Clarke (New York: Harper & Brothers, 1882).*

itself and all Polycleitus's original statues are now lost, two ancient sources survive that give us an idea of what Polycleitus had in mind (plate 5).

The first of these sources comes from the writings of the second-century CE Greek physician Galen. Recalling the third-century BCE Stoic philosopher Chrysippus, Galen wrote: "According to Chrysippus, beauty consists not in things generally matching but in the correspondence of the body's parts specifically: the commensurability, for example, of finger to finger, and of all the fingers with the metacarpal and carpal, and of these with the forearm, and the forearm with the upper arm, and of all of these with the entire body, just as Polycleitus says in his *Canon*."[11]

In his handbook of design and construction, the first-century BCE Roman architect and engineer Vitruvius had this to say:

No temple can possess a rational plan without symmetry and proportion. In other words, it must have the same precise plan as a well-designed human body. For Nature has so designed the human body that the head from the chin to the top of the forehead and the roots of the hair constitutes a tenth of the body's overall height; that the palm of the hand from the wrist to the tip of the middle finger measures the same; that the head from the chin to the very top of the head, an eighth; that the top of the chest including the bottom of the neck to the roots of the hair, a sixth; and the midpoint of the chest to the top of the head, a fourth. As for the face, the distance from the bottom of the chin to the bottom of the nostrils constitutes a third of its length, just as does the length of the nose from the bottom of the nostrils to the point between the eyebrows, as does the forehead from the point between the eyebrows to the roots of the hair. The other parts of the body are similarly proportioned. And by using these, ancient painters and noble sculptors have achieved infinite renown.[12]

Since both the face and the open hand of a perfect human specimen are, according to Vitruvius, each a tenth of its height and thus the same size, you can check to see if you are perfect by placing your hand over your face. If your hand and face are the same size, you, too, may be perfect—at least in the eyes of Vitruvius and possibly Polycleitus himself (fig. 30).

One remarkable by-product of this Polycleitan schematization was that most of the faces of fifth-century BCE statues—including that of Myron's *Discus Thrower* and the sculpted figures in Phidias's Parthenon Frieze—resemble each other. But in an age of idealism, the aim of Athens' sculptors was not to depict biological reality with all its idiosyncrasies. Instead, it was to create a patriotically inspiring image that transcended reality by identifying and representing the hidden essence of man, an essence proudly embodied in Athens' people.

In their effort to discover the "formula" for aesthetic and biological perfection, the sculptors of Athens' golden age were emulating the practice of those more ancient Pythagorean philosophers who had believed in an invisible, mathematically expressible order, or "cosmos," that resided in the heavens. Thus, in classical art, man was treated as a microcosm, a reflection in biological terms of the character and structure of the universe.

Vitruvius also went on in Hellenic fashion to describe the overall symmetry of the human body.

It naturally follows that the navel represents the body's exact center. For, were a man to lie flat on his back with his hands and feet outstretched and a circle were drawn using his navel as its center, its circumference would be

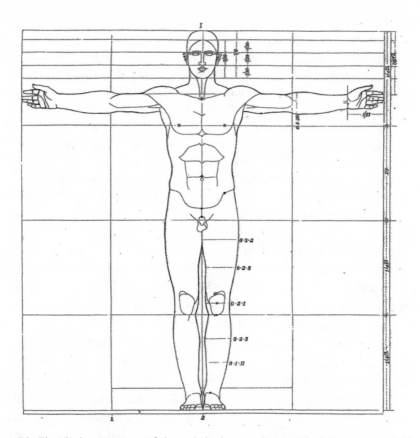

Figure 30: The ideal proportions of the male body according to Polycleitus as recorded in the writings of Vitruvius. George Redford, A Manual of Sculpture *(London: Sampson Low, Marston, Searle, and Rivington, 1882).*

touched by his fingers and toes. In similar fashion, a square could be drawn within that circle that would be tangent to his extremities. For if we were to measure from the soles of his feet to the top of his head, we would get the same distance as from one of his outstretched hands to the other, in much the same way that surveyors square a site.[13]

A millennium and a half later, this description would inspire Leonardo da Vinci to create his schematic and iconic drawing of man (see chapter 15). Furthermore, were Polycleitus alive today, he would marvel at today's attempts by genetic engineers to improve upon nature by manipulating DNA in order to produce more perfect bodies—not in lifeless bronze or marble but in living flesh.

THE BIRDS AND THE BEES

When, in the aftermath of the Peloponnesian War (431–404 BCE), the political ideals fostered by classical Athens disintegrated, Greek thinkers began to turn their attention once again to the tangible, nonhuman realities that surrounded them in the organic world of nature.

One of the greatest investigators of these realities was the fourth-century BCE philosopher Aristotle (384–322 BCE). Yet even Aristotle (fig. 31) revealed a humanistic bias, observing that man differed from all other creatures of the earth by virtue of the power of reasoning with which he, and not they, had been endowed.[14] Nevertheless, by activating his own rational powers, Aristotle turned to the exploration of the nonhuman world with unprecedented dedication, becoming in the process the "father of zoology." Indeed, over one-fifth of his extant works would deal with this subject. "Nature's every realm is filled with wonders," he wrote. "Therefore we should approach the study of every kind of living creature without hesitation or distaste, for each one will reveal something natural and beautiful."[15]

Figure 31: Bronze portrait of Aristotle, discovered in the ruins of Herculaneum and possibly dating to the fourth century BCE. From R. W. Livingstone, ed., The Legacy of Greece *(Oxford: Clarendon Press, 1921).*

Aristotle's quest for undiscovered beauty was matched by his conviction that a hidden order was latent in nature and could be discerned by man. It was the human intellect that could illuminate that natural, even divine design in which every creature and each of its anatomical parts had a distinct purpose.[16]

In the course of his research, Aristotle collected and examined specimens of some five hundred different species, including at least 60 types of insects and 120 kinds of fish, many of the latter caught in the lagoon of Pyrrha on the island of Lesbos, where he had honeymooned with his new bride before traveling to Athens to join Plato's Academy in Athens. Aristotle became the first Greek scientist to extensively perform dissections on animals, and he supplemented his own findings with valuable information gathered from fishermen and herdsmen, whose labors made them intimately familiar with creatures' habits.

Among the topics that fascinated Aristotle were the means by which animals (including humans) reproduce, how embryos develop, what environments and locales certain kinds of creatures inhabit, and what they depend upon for food. He believed, furthermore, that every biological organ had a specific function for which it was designed. The acute observations so characteristic of his work are illustrated by the following two extracts from his experimental research into the development of chicks and the habits of bees:

> By the tenth day, the whole chick and all of its parts are distinctly visible. Its head is still larger than the rest of its body, while its eyes are larger than its head, but are incapable of seeing. Around this time, the eyes become projecting, and are bigger than beans and black. When their skin is removed, there is a white and cold watery mass inside, which glistens in bright light but is no way solid. This then is the state of the eyes and the head.
>
> At this time, the internal organs are already distinguishable, both those that are associated with the stomach and those that are associated with the intestines. In addition, blood vessels can be seen extending from the heart to the navel. From the navel, one blood vessel extends toward the membrane surrounding the yolk (which is fluid at this time and larger than normal), while another extends to the membrane entirely surrounding (a) the membrane in which the chick lives, (b) the membrane around the yolk, and (c) the fluid between them. As the chick grows, little by little one part of the yoke moves up and one part down, while white fluid stays in the middle, the white of the egg being under the lower part of the yolk just as it was before. On the tenth day, the white is located at the extremities, and is small in quantity, and is sticky, thick, and very pale. . . .

Around the twentieth day, if you split the shell and touch the chick, it will make a sound as it moves inside. After the twentieth day when it breaks through its shell, the chick is already covered in down.[17]

The drones spend most of their time inside but, when they fly out, they rise up into the sky in a mass and circle round as though they were exercising hard. And, when they are done, they feast. The kings, however, [note: not the queens] whether for feeding or for any other reason, do not fly outside unless they are accompanied by the entire swarm. They say that, if a swarm goes astray, it will chase after its leader using its sense of smell until it finds him. It is said that, if the king is unable to fly, the swarm will carry him; and if he dies, the swarm will die. If the swarm outlives him for a time but does not make combs, no honey will be produced and the swarm itself will soon die.

Scrambling rapidly up the stalks of flowers, the bees pick up wax with their front legs, wipe them off on their middle legs, and then wipe those off on the bent parts of their hind legs. And when they are loaded, they fly off, visibly laden with the weight. On each flight the bee does not go to flowers differing in type but instead goes, for example, from violet to violet, not touching any other type. And, when it returns to the hive, it shakes itself off while being tended by three or four other bees. Exactly what these bees pick up [pollen] is not easy to see, and I have not been able to observe how they do their work.[18]

Perhaps Aristotle's single most influential contribution to biological science was his effort to uncover the organizational principle of the organic world by classifying all animals by type, thereby setting the precedent for the modern classification of living beings into their genus and species. For Aristotle, the major distinction in the animal world was between those that had blood, like humans, and those that did not—a biological distinction that reflects the essential difference between vertebrates and invertebrates. Though not arguing for evolution as Charles Darwin later would, Aristotle nevertheless identified an implicit "scale of being" or "ladder of life"; that is, a progression of organisms from the least to the most structurally complex.[19]

Aristotle's findings and conclusions on biology were summarized in three major works: *The History of* (i.e., *Inquiry into*) *Animals*, *On the Parts of Animals*, and *On the Generation of Animals*. In his endeavors he was aided by certain key traits of personality that are marks of rationalism and the hallmarks of the Hellenic mind: intense curiosity, a gift for detecting analogies, a masterful capacity for organization, keen analytical powers, and a talent for lucid exposition. Most importantly for science, Aristotle was committed to the search for truth, almost always letting the evidence speak for itself rather than manipulating the facts to make them fit a preconceived theory.

Of course, Aristotle was also human and he committed a number of errors, some (but not all) of them excusable because he was prevented by religious scruples from performing autopsies of human beings. Thus, he assumed that the heart, not the brain, was the seat of intelligence[20] and that the function of veins and arteries was essentially the same.[21] Most notably, he concluded that women had fewer teeth than men[22]—perhaps because he had not bothered to peer into Mrs. Aristotle's mouth, as British philosopher Bertrand Russell once wryly observed. In Aristotle's defense, recent research suggests that Aristotle may have been partially correct inasmuch as deficiencies of vitamins C and D in the diets of Mediterranean women may have led to calcium deficiency and tooth loss, especially when they were pregnant or lactating.

Nevertheless, despite these shortcomings, Aristotle would remain the unchallenged authority on zoology for two thousand years.

The Secret Garden

If Aristotle was the "father of zoology," his friend and fellow scientist Theophrastus (ca. 372–288 BCE) was the "father of botany."

Endowed with a mind as wide-ranging as Aristotle's own, Theophrastus authored some two hundred works on subjects as diverse as religion, psychology, and physics. Apart from his serious contributions to biology, his single most famous book is *Characters*, a set of thirty satiric profiles of annoying personality types, including the flatterer, the braggart, and the grouch.

In the field of botany, his fame rests on his fact-filled *History of* (i.e., *Inquiry into*) *Plants* and *Causes* (i.e., *Functions*) *of Plants*, volumes in which he classified varieties of vegetation (both those that thrived on land and in the sea), discussed how they grew and reproduced, and described their cultivation, in large part from his own experimentation and personal experience, thus paving the way for the painstaking studies of Gregor Mendel. In the preface to his work,

Theophrastus set forth the aim of his research in the following words: "We must consider the distinctive characters and the general nature of plants from the point of view of their morphology, their behavior under external conditions, their mode of generation, and the whole course of their life."[23]

Here are three brief extracts from that work:

The roots of every plant seem to grow before its above-ground parts, for a plant initially grows down into the earth. But no roots go farther down than the sun can reach since heat is responsible for their generation. Apart from this, the soil contributes in great measure to the growth of deep and especially long roots provided that it is light, loose, and porous.... Furthermore, younger plants that have reached maturity have deeper and longer roots than older plants because the roots of the older ones tend to wither away along with the rest of the plant.[24]

. . .

By and large, most annuals have two colors and two flowers. By "two flowers" I mean that one flower has another in its midst, as, for example, the rose, the lily, and the dark violet. Some annuals produce only a single true petal, but have others marked out, as it were, by lines, such as the bindweed. In the case of this flower, the petals are not separate. In similar fashion, the lower part of the narcissus has angular projections that extend from its edges. The same is generally true also of the olive tree's blossoms.[25]

. . .

With cultivation, the pomegranate and the almond can change: the pomegranate if it is cultivated with pig manure and plenty of water from a flowing stream; the almond if one inserts a tap and over the course of time draws off the sap as well as giving it whatever other care it may need. Likewise, it is clear that many wild plants can become domesticated, and many domesticated plants can become wild. For cultivation, on the one hand, and neglect, on the other, induce such changes. Some, however, would say that it is not so much a radical change as a natural progression from worse to better or from better to worse.[26]

Because Theophrastus lived in an age before chemical fertilizers and pesticides, his horticultural findings on interplanting, companion planting, and the effectiveness of different types of manure, remain invaluable to today's organic farmers.

Departing from Aristotle, Theophrastus challenged the notion that every natural event or organic part reflected intelligent design, though he shared with his mentor a mistaken belief in spontaneous generation—the belief that lifeless matter can give birth to living beings. Theophrastus also followed the folkloric practices of farmers in recommending that a gardener face east when taking roots out of the ground and eat garlic before harvesting certain plants.

Theophrastus and Aristotle were both alumni of Plato's Academy. When Aristotle founded his own school, the Lyceum, Theophrastus followed him and became his colleague. Before Aristotle died, he appointed his old friend to be the school's new dean, a position Theophrastus would occupy for thirty-five years. In addition to having made him the guardian of his two children, Aristotle had also bequeathed to him his library and personal papers and his treasured garden, a garden where Theophrastus the horticulturalist eventually chose to be buried at the age of eighty-five.

CADAVERS TO THE RESCUE

The career of Alexander the Great (356–323 BCE) and the Hellenistic age that dawned in the wake of his worldwide conquests had three major impacts on the development of biology. First, by expanding the geographic horizons of the Greeks, these two events opened their eyes to a new world that in turn stimulated their thinking with its exotic and multicultural wonders. Second, the founding of Alexandria's Library and Museum gave a grand new impetus to scientific research. And third, by attracting Greek colonists to Egypt, a land that had practiced the dissection of the dead as a sacred art for three thousand years, the inhibition long felt toward it by the Greeks was relaxed.

Two names came to be associated with the dissection of human cadavers in the name of science: the physicians Herophilus (330–260 BCE) and Erasistratus (ca. 315–240 BCE). Additionally, both performed vivisections on animals to better understand the physiology of a body that was alive. Indeed, dissections of human organs may have been performed as early as the fifth century BCE by Alcmaeon of Croton.

Herophilus made numerous anatomical discoveries. He identified the brain—not the heart, as Aristotle thought—as the operational center of the human nervous system, identified the brain's structural components, traced the circulatory network that brought it life-giving blood, and distinguished between the nervous system's sensory and motor nerves. In addition to inspecting the liver, he also explored the heart's main chambers, tracing the major blood vessels that led to and from them. He dissected the eye, peeling away and describing its distinct membranes, and located the ovaries of the female body, which he compared in function to the testes of the male. In the course of his investigations, he also coined many anatomical terms to name the body parts he had discovered. Adopted into Latin, these would forever become part of medicine's lexicon. The word *duodenum*, for example, comes from the Latin spelling for the Greek words for *twelve* and *fingers*, because, when Herophilus measured it with his hands, its average length equaled the width of twelve fingers laid side by side.

Herophilus's contemporary Erasistratus recognized that the heart's main function is that of a pump and that this human pump consists of four main "one-way" valves. He also deduced how food travels through the alimentary canal: propelled by muscular action. But, like many of his contemporaries, he thought that blood only traveled through veins, while arteries contained only air.

No greater testimony can be found to the need for persistence in scientific research than this passage from Erasistratus's writings:

> Those who are totally unfamiliar with research, once they begin to exercise their minds, become dumbfounded and immediately abandon the pursuit out of mental exhaustion, collapsing like runners who enter a race without prior conditioning. But the person who is experienced at research keeps trying every possible approach and every possible angle and, rather than giving up after a single day's labor, persists for the remainder of his life. Focusing on one idea after another that bears upon what he seeks, he presses on until he reaches his goal.[27]

During the days of the subsequent Roman Empire, medical researchers continued to practice dissections and even vivisections, perhaps encouraged by the supply of corpses generated by war, the brutality of slavery, and the public slaughter of both man and beast carried out in the name of mass entertainment. The most famous of biological researchers in those days was the physician Galen (129–199/216 CE). His experiments on live animals proved that, contrary to previous theories, both veins and arteries transport blood. In lieu of human cadavers, Galen operated on monkeys because of their anatomical resemblance to humans. Far from a sadist, he regarded the physiological insights he gained as a way of better ministering to his human patients.

Like many Greek scientists before him, Galen viewed science as a spiritual quest and the exploration of nature as a portal to the divine. In effect, the human body was a temple whose consummate and purposeful design revealed the mind of a master architect. His investigation of anatomy was, he said in humility, "a sacred book which I compose as a true hymn to him who created us: for I believe that true piety consists not in sacrificing many hecatombs of oxen to him or burning cassia and every kind of unguent, but in discovering first myself, and then showing to the rest of mankind, his wisdom, his power and his goodness."[28]

Though the Greeks were in essence humanists who took pride in their own powers, their scientists remained open to the possibility that there was a power beyond their own.

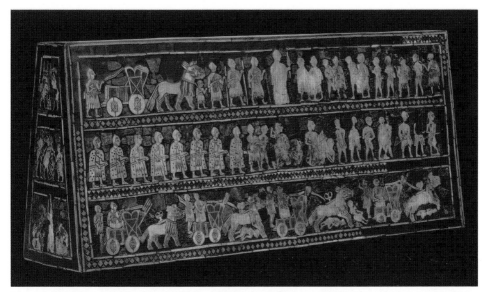

Plate 1: Military chariots depicted on the third-millennium BCE "Royal Standard" of Ur. *(Credit: © The Trustees of the British Museum; Art Resource, NY)*

Plate 2: View of the theater at Epidaurus. *(Credit: Vanni/Art Resource, NY)*

ARCHIMEDES erster erfinder scharpffsinniger vergleichung/
Wag vnd Gewicht/durch außfluß des Wassers.

Plate 3: Archimedes in his bathtub, from a sixteenth-century CE woodcut. *(Credit: HIP/ Art Resource, NY)*

Plate 4: Atlas holding the sphere of the heavens on his shoulders. National Archaeological Museum, Naples. *(Credit: Erich Lessing/Art Resource, NY)*

Plate 5: Georg Roemer's modern reconstruction of Polycleitus's original bronze *Spear-Bearer*. It was destroyed in Munich during World War II. *(Credit: Foto Marburg/Art Resource, NY)*

Plate 6: Major fragment of the "Antikythera mechanism." *(Courtesy of the National Archaeological Museum of Greece, Athens)*

Plate 7: Votive plaque showing a patient's shoulder being treated by the divine physician Amphiaraios *(left)* and a sacred serpent *(right)*. National Archaeological Museum of Greece, Athens. (*Credit: Erich Lessing/Art Resource, NY*)

Plate 8: The magnificent "Pont du Gard," a three-tier, 180-foot-tall aqueduct built in Roman France in the days of Augustus Caesar. *(Credit: Konrad Wothe/Photolibrary, NY)*

Plate 9: The death of Archimedes as depicted on an eighteenth-century copy of a Roman mosaic. *(Credit: Erich Lessing/Art Resource, NY)*

Plate 10: Sample page from the "Archimedes Palimpsest," with the original text and an original diagram still visible. *(Courtesy of the Walters Art Museum)*

Plate 11: "El Caracol," the astronomical observatory at Chichén Itzá, with the Temple of Kukulcán in the background. *(Credit: Leonardo Diaz Romero/Photolibrary, NY)*

Plate 12: Looking down the Street of the Dead toward the Pyramid of the Sun, Teotihuacán. *(Credit: Steve Vidler/Photolibrary, NY)*

Plate 13: Aerial view of Stonehenge and surrounding countryside. *(Credit: Adam Woolfitt/ Photolibrary, NY)*

Plate 14: View of Stonehenge with a trilithon framing the sun. *(Credit: Kasch/Photolibrary, NY)*

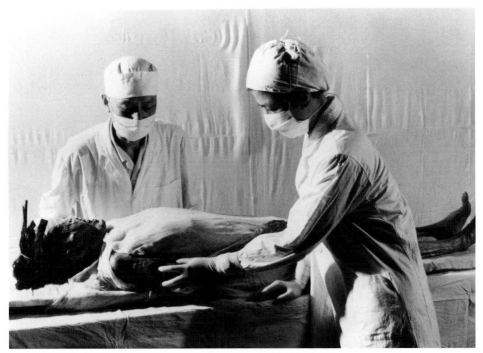

Plate 15: Autopsy being performed on the mummified "Lady in Red." *(Courtesy of Xinhua News Agency)*

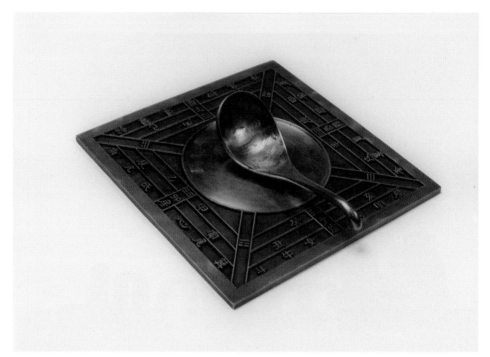

Plate 16: Modern reconstruction of an ancient Chinese lodestone compass. *(Courtesy of Science Centre, Toronto)*

Plate 17: Modern reconstruction of the Chinese "south-pointing" chariot. *(Courtesy of Science Museum, London/Science and Society Picture Library)*

Chapter 12

MEDICINE

Notwithstanding the fact that the art of healing had been practiced from time immemorial in the earliest civilizations of the ancient Near East, the Greeks were the first nation to turn that art into a true science by defining the issues of illness, treatment, and cure in rationalistic terms to a degree that was previously unprecedented. The ancient Egyptians had already made distinctions between spiritual remedies and medicinal substances, and the Mesopotamians had separate specialists skilled in the application of each, but neither culture elevated the role of the rational mind to effect physical well-being as much as the Greeks.

WARTIME MEDICINE

The annals of Western medicine began on the battlefields of Troy. In Homer's *Iliad*, the oldest-surviving work of European literature, battlefield injuries are graphically depicted (fig. 32). However, just recounting the names of the fallen heroes and where they fell was not enough for the epic poet or his audience. As a humanistic Greek with a scientist's eye for detail, Homer vividly portrayed the ways the injuries were inflicted and what specific part of a warrior's anatomy was affected. For example, in the fourth book of the *Iliad*, we read of a pitched battle between the Trojans and their Greek enemies:

> Stepping up to the front line, Simoisius was struck
> in his chest next to the right nipple. Straight through his shoulder
> the bronze spear went, and he fell to the ground
> in the dust like a toppled poplar . . .[1]
> A jagged rock hit Diores' right leg by the ankle . . .
> The pitiless stone crushed his two tendons

and the bones in his leg, and he fell backwards . . .
Piros then ran up and thrust his spear into
Diores' navel, spilling out his intestines onto the ground.[2]

Figure 32: Achilles binding a wound on the arm of his comrade Patroclus, depicted on the interior of an early fifth-century BCE Athenian wine cup. Berlin Museum. From R. W. Livingstone, ed., The Legacy of Greece *(Oxford: Clarendon Press, 1921).*

Fatal wounds were not the only ones described. Crossing the battlefield, the Greek warrior Patroclus ran into a comrade who had been hit in the thigh by an arrow and was limping off the field.

> *"Help me," he said, "and take me back to my ship.*
> *Then cut the arrow out of my thigh,*
> *wash the dark blood out with warm water,*
> *and sprinkle some soothing drugs on the wound,*
> *the kind they say Achilles instructed you to use,*
> *whom Chiron, the most righteous of the Centaurs, taught.*
> *For our medics, Podalirius and Machaon, are both unavailable:*
> *Podalirius lying in camp struck, I think, by a spear*
> *and in need of attention himself; and Machaon,*
> *caught up in the fierce battle against the Trojans on the plain."*[3]

From this passage we learn many things about the science of medicine in Homeric times. First, there *was* a science of medicine and it featured a treatment regimen and the use of pharmaceuticals (*pharmaka*, in Greek). Second, this science was employed in emergency situations by skilled and trained practitioners who risked their lives tending the wounded. And third, although it was a science, its principles were traced to the wisdom of mythic figures in Greek lore such as the half-man/half-horse Centaur named Chiron. Indeed, the physicians Podalirius and Machaon are spoken of as the sons of Asclepius, who would later be revered by the Greeks as their god of healing.

Significantly, the Greek word for physician—*iater*—is employed in the earliest inscriptions in Greek we possess: the so-called Linear B tablets that date to the very period, the Mycenaean, when Homer's heroes would have lived. From the standpoint of etymology, it is that very word, *iater*, from which we derive the familiar suffixes -*iatrist* and -*iatry* that denote medical specialists and specialties today.

Despite the fact that Homer's *Odyssey* takes place after the Trojan War, it, too, has much to tell us about the practice of medicine in early Greek times.

When Odysseus's son, Telemachus, visited the court of King Menelaus to gather news of his father's whereabouts, both he and Menelaus broke down in tears as they recalled the lost hero. To ease their anguish, Menelaus's queen, the beautiful Helen, "cast into the wine they drank a drug (*pharmakon*) that soothes sorrow and alleviates pain, and brings forgetfulness of every woe."[4] This drug and others, Homer tells us, was given to Helen by a woman named Polydamna when Helen and Menelaus had sojourned in Egypt, a land where "every man is a physician surpassing all other races in skill."[5]

Another psychotropic agent is described by Odysseus himself in the narration of his adventures.[6] The honey-sweet, flowery food called "lotus" induced memory loss in those who consumed it and deprived them of all motivation—except for the desire to keep eating it. Recognizing the addictive nature of this opiate and the threat it posed to the completion of his mission, Odysseus dragged his men, despite their tearful protests, back to their ships and immediately set sail.

The fact that the only drugs singled out in the poem have mind-altering effects further indicates the Hellenic fascination with the intellect at this early stage in Greek cultural history.

THE GOD OF HEALING

Notwithstanding their early rationalistic approach to the human body, the ancient Greeks also believed that cures had a spiritual dimension.

Their god of healing was Asclepius (fig. 33), later to be called Aesculapius by the Romans. In addition to being the father of Podalirius and Machaon, he was also said to have fathered Hygieia, who became a Greek goddess of health and, later, of mental health. It is from her name that the English words *hygiene* and *hygienic* come.

Figure 33: Statue of Asclepius, god of healing. Pergamum Museum, Berlin. From Harry Thurston Peck, Harper's Dictionary of Classical Literature and Antiquities *(New York: American Book Company, 1896).*

According to ancient tradition, Asclepius was the son of the god Apollo. Taken from his dead mother Coronis's womb, he was reared by the wise Centaur Chiron and tutored in the art of healing, with advanced training provided by Apollo himself.

Worshipped at some two hundred temples and shrines throughout the Greek world, Asclepius's most famous sanctuary was located in the southern Greek city of Epidaurus, a site renowned for its acoustically perfect theater. In ancient times, Epidaurus was the Hellenic equivalent of Lourdes; today the city hosts a drama festival enjoyed by throngs of tourists every summer.

Asclepius offered cures to patients whose conditions defied standard medical treatment. After making the necessary pilgrimage to his sanctuary, they underwent purification rites and offered small sacrificial cakes in his name, after which they entered an enclosed consecrated "sleeping hall," where they would undress and don white robes. Lying down on makeshift beds, the supplicants eventually fell asleep and, through a mystical process known as incubation, saw the god himself, who announced the way they could regain their health. The next morning, after receiving therapeutic advice from priests, the pilgrims made a donation to the temple treasury before departing, though they were also welcome to stay in the holy precinct for a "recovery" period, during which they might attend performances in the nearby theater.

Housed within Asclepius's sanctuary and slithering across its floor were non-poisonous snakes, sacred serpents that symbolized through the shedding of their skin the possibility of bodily renewal. Thus, the symbol of Asclepius became a staff entwined with a serpent. Ironically, when the modern medical community wanted a symbol of their profession, they chose such a staff, but it was surmounted by wings and entwined with not one but two serpents. Alas, this turned out to be not the classical symbol of Asclepius but the traditional symbol of Hermes, the Greek god of commerce. However, given the venal motives of some medical practitioners today, the choice may have had some validity.

The content of the dreams ancient patients had and the miraculous cures they received are described in inscriptions discovered in the sanctuary's ruins. Here are three such stories:

A boy who was mute came to the sanctuary to recover his power of speech. After the preliminary sacrifices and rituals, a priestly attendant asked the boy's father if he would swear to make an offering if his prayers were answered. Suddenly the boy said, "I promise!" Amazed, his father asked him to say it again. The boy did so and from that point on was cured.

A blind man came and was healed. But when he didn't give an offering to express his gratitude, the god made him blind again. Returning in repentance to the sanctuary, he was made well again.

During battle a man was wounded in his lung by an arrow. The wound became badly infected and, in the course of a year and a half, oozed huge quantities of pus. Having come to the sanctuary, he fell asleep and in his dream saw the god remove the arrow-point from his lung. When he woke up in the morning, the man was cured and held the point in his hand.

Similar testimonials are also found in a unique diary titled *Sacred Stories* kept by Publius Aelius Aristides, a Roman author of the second century CE. Plagued by chronic illness, Aristides made repeated visits to Asclepius's shrine at Pergamum and reported his dreams in detail and the cures he received.

Additional reports of so-called medical miracles survive in the form of terra-cotta models and marble plaques that were donated to the temple at Epidaurus and elsewhere by grateful patients. These images realistically replicated the body parts (such as limbs and organs) that were cured by the god's beneficent powers (fig. 34 and plate 7).

These testimonials and ex-votos constitute evidence that some type of healing did indeed take place. Among the conditions cataloged are arthritis, baldness, blindness, gallstones, gangrene, gout, infertility, infestations of lice, lameness, migraine headaches, paralysis, and tumors. Despite the skepticism we may harbor about the success of such healing, especially in the case of extreme examples, modern medical science has clearly established the existence of psychosomatic illnesses and, as evidenced through the documented placebo effect, has also revealed an undeniable therapeutic connection between mind and body. Furthermore, the notion that our dreams may hold the subconscious keys to unlock our maladies has also long been a central thesis of Freudian psychology. Significantly, Sigmund Freud himself was a student of antiquity and, as a reminder of its wisdom, kept numerous inspirational Egyptian and Greek statuettes in his study and on his desk.

A HUMOROUS APPROACH

While the military approach to medicine was mainly practical and the priestly approach spiritual, a theoretical approach to the question of disease came with

Figure 34: A grateful patient shown dedicating to the god Asclepius a votive image of his healed leg. From Arthur Fairbank, A Handbook of Greek Religion *(New York: American Book Company, 1910).*

the rise of the pre-Socratic philosophers in the sixth and early fifth centuries BCE. Following in the footsteps of Empedocles (ca. 492–432 BCE), who held that the material world consisted of four basic elements—earth, air, fire, and water, naturalistic Greek thinkers like Alcmaeon of Croton reasoned that the human body must also have four components, which they called blood, phlegm, choler (or yellow bile), and melancholy (literally "black" bile). As long as these "humors" (Latin for *fluids*) were in balance, a person was in perfect health (literally in "good humor"). But when there was an excess in the body of one or a deficiency of another, the result was pain and physical distress. Therapy, therefore, consisted of trying to restore the internal balance of these elements through diet and exercise, and by other means such as purging and draining excess fluids, including blood. The aim of restoring a healthy equilibrium in the body reflected the general Greek view that balance in all things is critically important, an attitude most famously expressed in the maxim "Nothing in excess" carved on the facade of the Temple of Apollo at Delphi.

Though the doctrine of humors was challenged in ancient Greek times by a number of thinkers who valued empirical evidence and experimentation over pure speculation, it would prevail until relatively modern times. It largely endured through the influence of the Greek philosopher and physician Galen, who lived and wrote during the days of the Roman Empire. And though it is no longer explicitly subscribed to by current medical science, the general notion of maintaining an internal balance within the body remains central to modern concepts of organic health. Furthermore, since an excess in each of the humors was thought to have a pronounced emotional effect, the words *phlegmatic, bilious,* and *melancholic* endure in the English language as descriptions of personality types and moods.

HIPPOCRATES U

By the classical age of Greece, a number of schools flourished as training grounds for physicians. The most famous was located on the Aegean island of Cos and was directed by a contemporary of Plato and Aristotle named Hippocrates (ca. 460–370 BCE), widely regarded as the "father of medicine." Some seventy medical textbooks, collectively known as the *Hippocratic Collection,* or *Corpus,* survive from that era, but whether Hippocrates was the author of some or all of them is hard to say.

Notwithstanding the fact that his school taught the doctrine of humors, extracts from the corpus reveal a keen sense of clinical observation, most evident in a series of actual case studies that detail on a day-by-day basis the specific symptoms presented by patients during the course of their illnesses. Rather than merely being historical records, the case studies served as guides that medical students and practitioners could use to predict the progress and probable outcome of a disease. The usefulness of such a prognosis is evident in the following passage from the *Collection.*

Forecasting is of immense importance to the physician. For if he can sit beside a patient and recount the entire history of the disease, not only describing what the patient has told him but filling in all the blanks, he will be more likely to win a person's trust. Moreover, he will have a better understanding of how he should treat a disease if he can predict its future direction.

It is frankly impossible to restore everyone to health. If it were, prognosis would count for less. Some men, in fact, simply die because they call the physician too late, their condition already being too serious or the physician lacking enough time to arrest it. . . .

Indeed the longer you prepare yourself to meet an emergency, the greater will be your power to save the patient; while you will be less likely to be blamed for their death if you can predict in advance who will live and who will die.[7]

It is notable that Hippocrates's name lives on in an oath sworn to by graduating medical students to this very day. According to the Hippocratic Oath, the physician swears to regard his teacher as his parent, to share his livelihood with him, and to educate his sons and his teacher's sons in the art of medicine without charge. He vows never to harm a patient, never to administer a deadly drug for the purposes of euthanasia, and never to induce an abortion. He also promises never to take sexual advantage of a patient or disclose a patient's confidences. All these things he swears to in the name of Apollo the Physician, Asclepius, Hygieia, Panacea (the goddess of all cures), and all the other gods. With this oath, Hippocrates showed that he regarded the craft of medicine as a sacred art, even though in the scientific treatment of disease he paradoxically discounted the role of supernatural forces at every turn.

The progress of Greek medicine would continue during the fourth century BCE and the succeeding Hellenistic age. Aristotle (384–322 BCE), the son of a physician to the royal Macedonian court, did not practice medicine himself, but, in systematizing what was known of the natural sciences in his day, he lent the precision of a master logician to his wide-ranging disquisitions on biology, anatomy, and physiology. As noted, however, he failed to distinguish between the functions of arteries and veins, and he regarded the heart rather than the brain as the seat of intelligence. Nevertheless, the distinction he and his teacher Plato made between the nature of the body and the nature of the soul helped to lift the centuries-long Greek ban on the dissection of corpses, opening the way to a deeper understanding of the workings of the human body and the treatment of disease. In the decades that followed, Greek medical researchers at Alexandria and elsewhere used cadavers to explore the circulatory, nervous, digestive, and muscular systems, and to study the body's internal organs, including the reproductive organs of both men and women.

In the fourth century BCE, Aristotle's colleague, Theophrastus, wrote a series of separate monographs on the physiology of fatigue, sweat, dizziness, fainting, paralysis, and bodily sensations. In addition, as a dedicated botanist, he investigated the medicinal properties of plants and the curative powers of perfumes.

Born in the first century BCE in far-off Bithynia on the Black Sea, a physician named Asclepiades proposed that the human body consisted of lumplike "corpuscles" that traveled through the body's ducts. Asclepiades also believed in applying noninvasive therapies like bathing, massage, and drinking wine and opposed the excessive use of drugs in his practice.

Although the impressive medical achievements of Galen (ca. 129–199/216 CE) belong chronologically to the glory days of Rome where he came to live as court physician to the emperor Marcus Aurelius, Galen was actually a Hellenistic Greek who received his training in the eastern Mediterranean, where he first worked as a physician tending gladiators. Perhaps because of that experience, he would later excel at surgery (fig. 35). Because of a sanction against performing dissections on human cadavers, Galen chose to operate on living animals in order to apply what he learned to more clearly understand the vital parts of the human body and their functions. Galen originally composed about five hundred separate treatises on all aspects of medical science, of which about a hundred survived the fall of Rome and exerted a lasting influence on Western thought (fig. 36).

Galen, in characteristically Greek fashion, regarded the study of philosophy as essential to the education of the physician:

> The treatise entitled *That the Best Doctor Is also a Philosopher* gives three main reasons for this. . . . First, the doctor must be trained in the scientific method. The emphasis here is not on the evaluation of evidence, so much as on knowledge of logic, the ability to set out a proof and to distinguish valid from invalid argument. Secondly, it is the task of philosophy to study nature, and the whole of the theoretical side of biology—the investigation of the constituent elements of the body and the functions of the organs, for example—comes under this heading. Thirdly, there is to us a rather surprising ethical reason for the doctor to study philosophy. The profit motive,

Galen says, is incompatible with a serious devotion to the art. The doctor must learn to despise money. Galen frequently accuses his colleagues of avarice and it is to defend the profession against this charge that he plays down the motive of financial gain in becoming a doctor. Just as we saw Vitruvius doing for architecture [see chapter 14], so Galen attempts as far as possible to assimilate medicine to philosophy, the supreme—because supremely unselfinterested—study.[8]

In addition to treatises on the practice of medicine, the Hellenistic age also produced compendia of pharmaceuticals. Probably the most popular of these ancient "Physicians' Desk References" was compiled by the first-century CE Greek researcher Dioscorides, who traveled throughout the eastern Mediterranean in search of herbs, minerals, and animal by-products that had medicinal effects. His travels yielded the names of about seven hundred different plants and a total of over a thousand drugs, which he painstakingly categorized by their sources and physiological reactions.

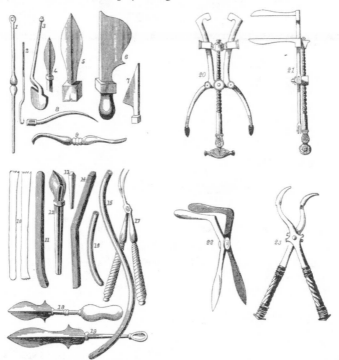

Figure 35: A variety of medical instruments, most found at Pompeii, used for physical examinations and surgery. From Harry Thurston Peck, Harper's Dictionary of Classical Literature and Antiquities *(New York: American Book Company, 1896).*

Figure 36: Tombstone of a second-century CE doctor named Jason, portrayed treating one of his patients. A cupping device, used to draw blood to the surface of the skin, can be seen at the lower right. From R. W. Livingtone, ed., The Legacy of Greece *(Oxford: Clarendon Press, 1921).*

AN ANATOMY OF DISASTER

Strikingly, one of the most salient demonstrations of the Greeks' scientific attitude toward disease came not from the writings of a physician or pharmacologist but from the diary of a military leader and historian.

His name was Thucydides, and he was a participant in the great Peloponnesian War between Athens and Sparta. Given command of a naval unit and unjustly blamed for a defeat, Thucydides was sentenced to exile from Athens. Now separated from his home city, he set about writing a history of the war and its causes. In his account, he accused his fellow citizens of hubris. Their blind arrogance, he argued, had spawned imperialistic policies that inevitably led to Athens' fall and to the end of its idealistic golden age.

In one of the most powerful episodes in his history, Thucydides described a plague that struck Athens at the very time its population, under siege by the Spartans, was penned up inside the city walls.[9] Thucydides detailed the symptoms and progression of the deadly epidemic with a clinical precision and detachment that is extraordinary, considering he himself fell victim to the disease. Thucydides would survive. Epidemiologists today still debate the precise agent of the outbreak. Its symptoms suggest not bubonic plague but possibly typhus or perhaps even a precursor of the Ebola virus.

For no apparent reason, those who had been healthy suddenly developed a severe fever in their heads and a redness and inflammation in their eyes, while internally their tongues and throats became blood-red and their breath peculiar and foul. Next came sneezing and hoarseness, followed soon by chest congestion marked by a deep cough. Settling into the stomach, the sickness then upset it, inducing persistent vomiting (in the various colors known to physicians) accompanied by exhaustion. An ineffectual retching affected the majority of sufferers, leading to violent spasms that occurred for some right after the previous symptoms had abated, but for others a long time thereafter. Their skin was not especially warm to the touch nor did it look pale, but was instead reddish or livid, and covered with small pustules and sores. The victims were so hot they couldn't stand having their bodies covered by even the lightest of clothing or sheets but wanted instead to be naked and, especially, to plunge into cold water. Many who were unattended even ran to the public cisterns, experiencing a thirst that was not slaked no matter how much they drank. Throughout, they were oppressed by restlessness and insomnia. And the body, as long as the disease was in full course, did not waste away, but improbably held up under extreme hardship, so that most expired on the seventh or ninth day from the intense fever while still retaining some energy. If, however, they survived this, most later died from exhaustion when the disease went down into the bowel, causing severe ulceration and incessant diarrhea.

Thus the malady, having first started in the head, went down through the whole body. And if someone escaped the worst, it left its mark by attacking and destroying the extremities—the genitalia, the fingers, the toes, and even the eyes. Immediately upon their recovery, some experienced total amnesia and knew neither their friends nor their very own selves.[10]

Being a humanistic Greek, Thucydides was fascinated less by the epidemic itself than by how people behaved when they knew a death sentence was hanging over their heads. As we see from the scathing report below, those who had once been portrayed in the marble sculptures of the Parthenon as nearly godlike soon proved themselves to be worse than beasts.

Dead bodies were piled on each other or were sprawled in the streets, while bodies lay around the fountains half-dead with thirst. The shrines that some had gone to live in were packed with the corpses of those who had already died there. For when the disaster overwhelmed them, people, not knowing what would come next, showed no regard for whether something was sacred or profane. All the laws governing burial that once were observed were upended and each buried his dead where he could. For want of supplies due to the multitude of their loved ones who had already perished, many disposed of their newly deceased relatives shamelessly. Some, who happened upon another person's pyre built by his family, laid their own corpse upon it and lit the fire, while others tossed their body onto an already burning pyre and ran off.

Besides this, the epidemic introduced a greater degree of lawlessness into the city than had ever existed before. People dared to do for pleasure openly what they had once done only in secret given the abrupt change they witnessed as the rich suddenly died and the poor inherited their wealth. And so they did whatever could provide them with instant gratification and enjoyment seeing that their lives and possessions were but things of a day. People were unwilling to sacrifice for what was honorable inasmuch as it was unclear whether they would ever survive to enjoy the reputation for it. Instead, the pleasure of the moment and everything that could contribute to it was deemed good and proper. No one feared god's law or man's but, seeing as everyone perished equally, saw no difference between respecting the law or not. No one expected to be alive to face a trial for his crimes; rather a far greater sentence had already been pronounced and hung over his head, and before it was carried out it seemed only reasonable to get something out of life.[11]

Thucydides told us he compiled this history not for mere entertainment, as had his predecessor Herodotus, but with deadly serious intent.[12] Like his contemporary Hippocrates, Thucydides was convinced of the value of prognosis, believing that if future readers could comprehend the errant steps that had led a former nation from grandeur to disaster, they could save themselves from a similar fate. Thus, he argued, the study of history may be the most potent tonic a people can take to ensure their civic health.

A second example of diagnosis from a nonphysician is even more extraordinary because the patient in question was dispassionately reporting the progressive stages of his own death. The dying man was Socrates (fig. 37), who by Athenian court order was forced to drink a potion of deadly hemlock juice five years after the city fell to Sparta. Here are his closing moments, narrated by his disciple Phaedo. The year is 399 BCE; the setting, an Athenian prison.

> So saying, putting the cup to his lips, he drank the poison down calmly and without hesitation.
>   Up until then most of us had done a pretty good job of holding back our tears, but when we saw him start drinking and then finish, we couldn't

Figure 37: Socrates, after a sketch by Peter Paul Rubens. From J. P. Mahaffy et al., The Classics Greek and Latin, *vol. 4, "Philosophy" (New York: Vincent Parke, 1909).*

do so any longer and, in spite of myself, my tears gushed out and, hiding my face, I sobbed uncontrollably, weeping not so much for him as for myself in being deprived of such a dear friend.

Crito, in fact, had gotten up before me because he couldn't hold back his tears either. But Apollodorus, who had been crying nonstop all the while, burst out in loud and anguished sobbing, making all of us who were present break down—all except Socrates himself who said, "What in the world are you people doing? This is the very reason why I sent the women away, so they wouldn't make such noise, since I've heard that it's best to die in solemn silence. So be still and hang on." When we heard those words we felt ashamed and held back our tears.

Once he had walked around, he said his legs were getting heavy, and so laid down on his back as the attendant suggested. The man who had administered the poison felt Socrates and then, after letting some time go by, examined his feet and legs. Pinching Socrates' foot hard, he asked him if he felt it. He said, "No." After that, when his thighs became numb as the effects of the poison moved upwards, he showed us how Socrates' body was becoming progressively more cold and stiff. He touched him once again and said that, when it reached his heart, Socrates would pass away.

When the chill had reached his abdomen, Socrates, who had up to that point covered himself up, uncovered himself and spoke these dying words. "Crito," he said, "we owe Asclepius a rooster. Make sure the debt is paid." "It'll be taken care of," said Crito. "Is there anything else you want to say?" He didn't respond to the question but, after a little while, moved. The attendant then uncovered him and his eyes were motionless. When Crito saw this, he closed Socrates' mouth and eyes.

Such was the end, Echecrates, of our friend, a man, we could say, who was the best of those who lived in those days and, besides, the wisest and most just.[13]

The rooster Socrates wanted sacrificed to the god of healing is thought to have been intended to restore the health of his disciple Plato, who was then absent because of illness. Despite Socrates's rationalism, he still believed in honoring the gods.

Chapter 13

PSYCHOLOGY

We are indebted to the ancient Greeks for the science of psychology, the rational exploration of the workings of the human mind. As a race of explorers, it is only natural that the Greeks should have invented psychology, for the most fascinating thing they could think of to explore as humanists was themselves. In fact, in some ways the practice of psychology represented to them the ultimate intellectual challenge: to turn the mind upon itself to see what it could find in its own inner recesses. So pervasive was their fascination that psychology became a dominant theme of their literary tradition.

EPIC EMOTIONS

The earliest evidence of this theme occurs in the oldest extant work of Greek literature, Homer's *Iliad*. In the epic poem's opening verses, indeed with its very first word in Greek, the poet declares that his plot will be driven by the darkest of emotions: *menis*, or wrath.

> Let wrath be your theme, o Goddess, Achilles'
> deadly wrath, that caused immeasurable suffering
> to the Achaeans, and hurled into Hell the valiant souls of
> heroes, leaving their corpses as carrion for
> dogs and birds, thus fulfilling the will of Zeus.[1]

Achilles, we soon learn, had come to fight for glory in the Trojan War, believing that such glory would bestow upon his name an immortality that he was denied by life itself, having been born of a goddess-mother but an all too mortal father. For nine years the war had dragged on, and for nine years Achilles had been demeaned by his commander-in-chief, Agamemnon. In a final act of disrespect, Agamemnon had wrested Achilles's prized concubine from his arms. In retaliation, Achilles withdrew from the war and withheld the men he commanded, an act that would almost certainly spell the Greeks' defeat.

Seething over the insults to his pride and torn by his own desperate decision to abort his search for glory, Achilles sat alone on the beach, looking out over the wine-dark sea, and wept. It was then that his divine mother, the water-nymph Thetis, hearing him cry, arose from her home in the depths of the sea to bring him comfort. "My child," she asked, "why do you weep? What sorrow has overtaken your heart? Don't bury it in your mind, but speak out so both of us may know."[2]

Here, in the oldest-surviving work of Western literature, we find the beginnings of psychotherapy. The Greek bard recognized, and his ancient audience understood, that only by unburdening ourselves of our inner torment by articulating our pain can we heal our aching soul. And only by sharing our pain with another whom we trust can the full weight of that pain be lifted.

Although the solace Thetis brought her son gave him a temporary respite, Achilles's decision to avenge himself on Agamemnon by absenting himself from battle would cost the lives of hundreds of his fellow warriors who had depended upon him to fight beside them. When Achilles's compassionate friend Patroclus saw what was happening and chose to aid the Greek cause by fighting in Achilles's stead, he was killed. Crushed by his best friend's death, Achilles then decided to divert his wrath from Agamemnon and direct it instead toward Patroclus's killer, Hector, the leading defender of Troy. In a savage onslaught, Achilles cut down the Trojans in his path and slew Hector, venting his wrath on his fallen enemy by tying his corpse behind his chariot and dragging it over and over around Troy's walls before the eyes of Hector's anguished parents.

But mutilating the body of the man who had slain his best friend, Achilles discovered, would neither bring his best friend back to life nor give him satisfaction. There are some hurts, the poet revealed, that cannot be undone. Achilles, moreover, never introspectively acknowledged his own complicity in his friend's death, for if he had never withdrawn from battle himself and had never let his friend act as his surrogate, Patroclus would still be alive.

In a poignant scene at the end of the epic, Hector's aged father Priam, the king of Troy, comes to beg Achilles for the body of his son so he can give it proper burial and permit its soul to rest. Reminded that his own father may never see him alive again, Achilles empathetically relents and grants Priam his wish, but not before his wrath flares again. Pressed to release the body as soon as possible, Achilles threatens to kill the old man on the spot. Soon after Priam leaves, the poem ends, but the deadly war inexorably goes on.

The *Iliad* is a tragic tale because it offers neither final reconciliation nor the prospect of personal growth. Instead, it reveals how destructively intransigent our emotions can be and how blind we can be to the ways they subconsciously control our lives.

Notwithstanding their historic reliance on rationality, the ancient Greeks were not creatures of pure reason. We would be seriously mistaken to assume they were. Instead, "Don't go to extremes" (*Meden agan*) was inscribed on the facade of Apollo's temple at Delphi precisely because the Greeks, for centuries after Achilles, continued to be a passionate people habitually prone to excess. Were they not, such a warning sign would not have been needed. Perhaps it is our modern familiarity with Greek statues, sculpted from the coolest and purest white marble, that leads us to the false conclusion that the Greeks themselves were not hot-blooded. Yet what we fail to see today are the bright colors with which the now anemic marble was once tinted to make it look truer to life, for the lives of the ancient Greeks were anything but colorless. Similarly, their preoccupation with philosophy may delude us into thinking they had achieved ultimate self-mastery, whereas in fact they persisted in that search precisely because the final goal continued to elude them like a horizon that perpetually recedes before a traveler's advance. Remember too: all those old philosophers whose bearded busts are now preserved in dusty museums began their lives as young men vitally yearning for answers.

THEATER AS THERAPY

Carved high above the doorway to the Temple of Apollo at Delphi were these words: *Gnothe seauton.* "Know thyself!" It is the hardest task we humans can ever attempt to accomplish, the Greeks believed, yet there is no task more vital.

It is for this reason that the playwrights of classical Athens wrote tragedies, which were performed at holy festivals every year. The playwrights viewed themselves as teachers, and the theater was their schoolroom. As Aristotle noted,[3] drama invites the members of an audience to enter into the lives of characters on a stage and, by identifying with them and their torment, leave the theater purged—at least temporarily—of the dark impulses they witnessed.

Of the authors of classical tragedy, the most renowned were Aeschylus, Sophocles, and Euripides. According to Sigmund Freud, long before the unconscious was systematically investigated by psychoanalysts, it had already been discovered by storytellers: "Story-tellers are valuable allies, and their testimony is to be rated high, for they usually know many things between heaven and earth that our academic wisdom does not even dream of. In psychic knowledge, indeed, they are far ahead of us ordinary people, because they draw from sources that we have not yet made accessible for science."[4]

The Greek playwright Aeschylus explored the destructive consequences of willfulness and pride. His prime example was Agamemnon, the king of Mycenae. Agamemnon was the commander-in-chief of the Greek army that set out in a thousand ships to attack Troy, retrieve the beautiful Helen, and plunder the city's treasure. Storm winds, however, prevented the armada from setting sail from the Greek coast. To placate the goddess Artemis and secure favorable winds for a successful voyage, Agamemnon consented to offer up his daughter Iphigenia as a human sacrifice rather than suffer the humiliation of turning back. Agamemnon's wife, Clytemnestra, would never forgive him for this fateful decision. She took a lover who had a score of his own to settle, and waited for her husband's homecoming to exact her revenge.

At the war's end, Agamemnon returned, brazenly accompanied by a Trojan concubine. Rolling out a purple carpet to test his vanity, Clytemnestra invited him to step down from his chariot. Protesting that such finery befitted a god rather than a mortal, Agamemnon nevertheless acceded to her blandishments even as the chorus, witnessing his act, was filled with a strange foreboding. Inside the palace, as Agamemnon took a relaxing bath, Clytemnestra threw a net over him and stabbed him to death, exulting in her triumph. Later, in seeking to avenge his father's murder, Agamemnon's son Orestes would slay both his mother and her lover and be hounded in his guilt by the avenging Furies for the shedding of parental blood.

Aeschylus's violent play explores the deadly consequences of self-absorption and wanton arrogance, or *hubris*, an attitude that recognizes no limits beyond its own gratification. Because calling off the Trojan War would have frustrated and stifled his ambition and diminished his standing in the eyes of other kings, Agamemnon chose to shed his own daughter's blood. Having done this, and having succeeded in his mission to take Troy, Agamemnon then had the effrontery to return home in the company of a mistress. He paid the price. As a race of achievers who relished the exhilaration of success, the ancient Greeks were peculiarly prone to hubristic excess. Though they intellectually recognized its dangers (and even depicted them in drama), they were often unable both individually and nationally to resist its addictive power.

Our next playwright, Sophocles, argued that what happens in our lives is the product of a confluence between character and circumstance—yet neither character nor circumstance is within our full control.

Having learned from the oracle of Apollo at Delphi that he would murder his father and marry his mother if he returned home, a young man named Oedipus headed in the opposite direction. Coming to a narrow defile on the road to Thebes, he encountered a crusty old man in a carriage who refused to make way. Enraged, Oedipus struck out and killed the old man, along with the attendants who accompanied him. Continuing down the road, Oedipus next encountered a monster who barred the way. "Answer my riddle," the Sphinx shrieked, "and I will let you pass. Fail to answer it and you will die!" The riddle was this: "What walks on four at sunrise, on two at noon, and on three at sunset?"

Oedipus, who was as nimble with his wits as he was quick with his fists, replied, "Man, for at the dawn of his life he crawls on all fours as an infant, at noon walks upright on two legs as an adult, and at the sunset of his life totters on two legs and cane." It was the correct answer, and in consternation the Sphinx promptly and conveniently hurled herself off a cliff.

Proceeding down the road, Oedipus arrived at the city of Thebes, whose citizens joyously welcomed him as a liberator, since he had dispatched the monster that had long threatened their community. In gratitude, they offered him the hand of their recently widowed queen, an offer that ambitious Oedipus swiftly and gratefully accepted.

Some years later, however, a plague struck Thebes. Believing it was punishment for some great sin, Oedipus appealed to the oracle at Delphi to learn what he might do to placate the gods and bring relief to the citizens of his newfound home. "Find the man who murdered the king," he was told. Drawing upon all his intellectual resources, Oedipus turned into a detective, compulsively determined to apprehend the murderer at all costs. Eventually, Detective Oedipus tracked down his quarry only to discover the murderer was himself. The crusty old man he had killed on the road to Thebes had been its king. More than that, the king had been his birth father, for the Corinthian couple Oedipus had always thought of as his parents had actually adopted him as an infant. More horrible still (as Sigmund Freud would point out twenty-five centuries later), the queen Oedipus had married and slept with was his real mother. In anguish at this revelation and laden with guilt (his mother would hang herself when she found out), Oedipus blinded himself with the queen's sharp broaches and left the city in shame.

The answer to the Sphinx's riddle had not only been "man" but Oedipus too: the baby who was adopted at sunrise, who strode confidently on two legs at noon, and who now—in the sunset of his career—hobbled into exile on a cane. But the riddle of man has more to tell us. As Sophocles saw, man is a creature undone not only by his emotional weaknesses but even by his intellectual strengths. In the play, it is a blind prophet, Tiresias, who sees the truth, a truth to which the sighted Oedipus is ironically blind.

Of all the psychologists of the Athenian theater, however, the most penetrating in his analysis was Euripides. Rather than view the characters of his dramas as larger-than-life figures from myth, Euripides brought them down to Earth and portrayed them as human beings governed by emotions and needs that his audience could readily recognize in themselves. In his tragedies *Medea* and *Hippolytus*, he examined with almost surgical precision how revenge could be born of love.

Medea, an exotic sorceress from the land of Colchis, met the Greek hero Jason on his quest to seize her kingdom's treasure, a fleece of pure gold. Falling hopelessly in love with the stranger, she aided him in his quest, betraying her father and slaughtering her young brother so Jason could complete his mission. Returning with him to Greece, she took up residence in Corinth and bore him two sons. But Jason, eager to advance his own career, spurned the woman who had helped him and chose Corinth's princess for his new bride. Scorned and threatened with exile, Medea planned her revenge. She presented her rival with a gown secretly smeared with a corrosive poison and, to make her revenge on Jason complete, slew her two sons with her own hand.

Phaedra, the Cretan-born queen of Athens, proved just as vengeful in Euripides's *Hippolytus*, but for a different reason. Against her will, Phaedra fell madly in love with her stepson Hippolytus, who valued his virginity above all other things. When Hippolytus callously rejected Phaedra's sexual advances, she committed suicide, leaving behind a note that accused him of having forced himself upon her. Finding the note, an enraged King Theseus called down a deadly curse upon his son's head, only to learn too late that Hippolytus had been innocent all along. In this play as in *Medea*, Euripides reveals how erotic passion can overwhelm us and lead to tragedy, mindless as we are of how it can ravage our lives. Yet the playwright also demonstrates how totally renouncing sex can be equally destructive. Instead, we must avoid extremes, paying equal homage both to Artemis, the goddess of chastity, and to Aphrodite, the goddess of love.

Like his fellow psychologists of the stage, Aeschylus and Sophocles, Euripides depicted how tragically oblivious we can be to the dark forces that dictate our behavior. Whether driven by ambition like Agamemnon, Oedipus, and Jason, or by revenge like Clytemnestra, Medea, and Phaedra, we unconsciously write the closing act of our own lives. As a later dramatist, William Shakespeare, would wisely note: "The fault is not . . . in our stars, / But in ourselves."[5]

Perhaps the greatest tragedy of all was that the Athenians themselves did not heed the warnings of their playwrights. Ignoring the inevitable cycle that leads from affluence (*olbos*) to arrogance (*hubris*) to blind folly (*ate*) and ultimately to divine vengeance (*nemesis*), they built an empire in the classical age that came crashing down around their heads. As the cool and dispassionate faces of their statues reveal, the supremely self-confident Athenians had convinced themselves that they were in total control of their own destiny. In their case, as in the case of a host of other tragic heroes and heroines, self-knowledge would come too late.

THE SCIENCE OF LOVE

As we continue our psychological journey, let us return to the subject of erotic passion and to its reigning goddess, Aphrodite (figs. 38 and 39).

The love poems of ancient Greece are the oldest in Western literature. Besides being masterpieces of poetry, they are also extraordinary documents of personal introspection that reveal the capacity of the ancient Greeks to analyze the ecstasy and agony of their sexual experiences.

Figure 38: Bronze portrait of the goddess of love, Aphrodite. British Museum. From George Redford, A Manual of Sculpture *(London: Sampson Low, Marston, Searle, and Rivington, 1882).*

Figure 39: The Venus de Milo, *a portrait of Aphrodite discovered on the Greek island of Melos and now in the Louvre. From Max Sauerlandt,* Griechische Bildwerke *(Dusseldorf and Leipzig: Karl Robert Langewiesche, 1907).*

Following are but three examples; however, the testimony of each is compelling.

The first is a poem by the sixth-century BCE poet Ibycus, in which he contrasts the tranquil outer world of nature with the raging inner universe of human emotion.

> *Spring has come*
> *and quinces and pomegranates drink*
> *from the streams*
> *where the nymphs' pure orchard is,*
> *and vine flowers grow beneath tendrils*
> *shaded in flourishing leaves.*
> *But for me*
> *love has no season of repose.*
> *For flashing into fire,*
> *Thrace's north wind of passion,*
> *rushing with scorching*
> *fury, dark, fearless,*
> *grabs at the roots and shakes*
> *my heart.*[6]

The next poem—on the subject of erectile dysfunction—dates to the first century BCE and is by a poet named Philodemus.

> *I used to come five or ten times in one night,*
> *but then I couldn't get it up once, try as I might.*
> *Little by little this damned thing is dying*
> *(not that it wasn't already half-dead),*
> *and the way things now seem to be going*
> *I'll pretty soon be losing my head.*
> *Considering the current state of my sexual woes,*
> *imagine what Old Age'll do when he finally shows!*[7]

Our last selection comes from the second century BCE. The verses were found scribbled on a crumbling piece of papyrus near Alexandria, Egypt. The name of the author, a woman, did not survive.

The two of us had a choice,
and we chose to be together,
putting our faith in Aphrodite.
Whenever I remember
how he kissed me, intending all the while
to leave, causing confusion, I hurt.
Yet once affection became desire,
I could not banish him from my mind.
Dear Stars and divine Night, partner in my love,
convey me even now to him
whom Aphrodite made my master,
Aphrodite and the passion that seized my soul.
For a light to guide me
I have the blazing fire in my heart.
The fire's my torment. The fire's my pain.
That liar, who had the arrogance to deny
true love ever brought us together,
he is the one truly at fault.
And I, I'm going mad,
possessed by jealousy,
consumed by rejection.
Just throw me some flowers
to color my loneliness with.
My lord, don't send me away.
Take in one who is now shut out,
for I will gladly be your abject slave,
so desperate am I to see you.
(The hard thing is to be strong and tough.
They say if you give your heart to one man,
you'll lose your mind in the end,
for love that is so narrow
can only end in madness.)
I'm not one to quit when
trouble comes my way,
but it drives me mad
to think I lie alone
when you're out

playing the field.
So let's call a truce
and make up.
Isn't that why we have friends,
to tell us who's been wrong?[8]

DRUNKEN CENTAURS

Greek mythology is layered with psychological insights, and one of the most revealing myths concerns the battle of the Lapiths and the Centaurs.

Once upon a time, so the story goes, the king of the Lapiths decided to get married and have a big wedding to which all the neighboring tribes would be invited. Among those tribes was a peculiar one whose members had the bodies of human beings from the waist up but the bodies of horses from the waist down.

Soon after arriving at the reception, the Centaurs drank too much and got drunk. Ogling the bride and the bridesmaids, they decided to snatch them up and gallop off. When they made their move, however, the Lapith men defended their women and, after a furious fight, drove the Centaurs off (fig. 40).

To the mind of the ancient Greeks, always intrigued by human psychology, this mythical episode was fraught with deeper meaning. The Centaurs, they noted, were only half-human and, as such, subject to their animalistic impulses. Drinking too much wine freed them of their inhibitions, and they acted out their desires. The Lapiths, on the other hand, were fully human, and so the hand of reason checked their impulses. To the Greeks, the Lapiths represented the definition of civilized man; the Centaurs, barbarism. As Aristotle notes, it is the faculty of reason that distinguishes human beings from animals.[9] In classical Athens, this mythic struggle between civilization and barbarism was illustrated in battle scenes on the metopes of the Parthenon and in the contrast between frenetic horses and their dispassionate riders on the Parthenon's marble frieze.

As the Greeks would be quick to point out, the battle of the Lapiths and the Centaurs still rages on in each of our souls. Will we let our impulses (our *id*, as Freud would say) govern our lives or will we restrain those impulses under the guidance of moral tradition and law (Freud's *superego*)?

Figure 40: Struggle between a Centaur (left) and a Lapith (right) on a metope from the Parthenon. From Ernest A. Gardner, Six Greek Sculptors (London: Duckworth, 1910).

THE LAND OF THE LOTUS-EATERS

The *Odyssey*, the companion poem to Homer's *Iliad*, tells of the homeward adventures of Odysseus following the conclusion of the Trojan War.

One of the first of these adventures takes place in the tropical land of the Lotus-Eaters. When Odysseus sends out a scouting party, its members fail to return. As described earlier, Odysseus later discovers them in the company of friendly natives, sharing with them a sweet fruit called the lotus. The problem is that eating the fruit is so pleasurable it robs one of any thought of doing anything else, including sailing home. Recognizing the danger, Odysseus drags his men back to their ship, ties them up (lest they jump overboard and swim back), and immediately sets sail.

The episode of the Lotus-Eaters shows how dangerous sensual pleasure can be, but it is not the only such episode in the poem. When Odysseus's men naively accept food and drink from the enchantress Circe, they are instantly transformed into pigs (as befits their piggish behavior). And when Odysseus is held prisoner by the beautiful goddess Calypso on her island paradise, he is forced to choose between an eternity of sensual delight in her company or a mortal life of responsibility and struggle back home in Ithaca. The dichotomy between self-gratification and self-sacrifice could not be clearer.

PRISONERS OF THE CAVE

The nature of the senses became an important issue for many Greek philosophers of the fifth and fourth centuries BCE, including Anaxagoras, Empedocles, Democritus, Plato, Aristotle, and Theophrastus. The focus of their concern was causation, the mechanics by which we see, hear, smell, taste, and feel. In addition, they examined such special topics as memory and illusion, the origin of dreams, and the effects of music, drugs, and age on the mind. Aristotle (384–322 BCE), for example, theorized that the lingering impressions of past experiences take visual form during sleep.[10] He also observed how youth and old age are characterized by diametrically opposed emotional states, with the young tending toward idealism and impulsiveness, and the elderly exhibiting practicality and caution.[11]

On the other hand, Plato (ca. 429–327 BCE) examined the senses from a moral rather than a purely physiological perspective. In so doing, he was reacting to the fall of Athens and the delusions of grandeur, elsewhere delineated by the historian Thucydides, which had been responsible for Athens' imperialistic adventures and moral degeneration.

In the Allegory of the Cave[12] Plato, as noted earlier, describes a group of prisoners permanently chained to the floor of a dark cave who for all their lives had only been permitted to see shadowy images artificially projected on the wall in front of them. Eventually, one prisoner manages to break free of his shackles and stumbles his way up the incline to the cave's mouth, from which he steps out into the light of day. At first dazzled by the intense light of the sun, his eyes gradually adapt to its brilliance and begin to discern objects from the natural world that they had never before beheld. Realizing how his lifelong indoctrination had blinded him to physical reality, he descends once again into the darkness of his former prison in a vain attempt to convince his fellow prisoners that they, too, must escape.

The cave, Plato explains, is a metaphor for the darkness in which we are all entrapped by the society into which we are born, a society that educates us in its false and shallow constructs. We must break free of convention and struggle to see reality for what it truly is. Indeed, no longer depending on our senses that offer only a counterfeit of reality, we must exercise our minds until, by a long process of re-education (so very different from today's self-help quick fixes), we can see a higher truth and eventually lead others to it. This is the task of the philosopher who, by long training and self-discipline, becomes able to look directly at the sun itself, the ultimate reality and symbol of the Good, by whose ethical rays the rest of the world is made visible. Nevertheless, recalling the fate of his mentor Socrates, who was executed by his beloved Athens for forcing people to face their illusions, Plato admits that attempting to lead others to enlightenment will always be fraught with danger.

THE TRINITY OF THE SOUL

Developing the distinction between sensory illusion and reality, Plato proposed that the human personality is a hierarchy composed of three parts.[13] The lowest and basest of these—a part we have in common with animals— is made up of the appetites, the primal urges—for food, drink, and sex—that are rooted in the senses. The second part consists of will, the sum of the emotional drives—such as love, anger, patriotism, and ambition—that energize our actions. Third, and highest of all, is the intellect, which can distinguish right from wrong and good from bad.

For a personality to thrive there must be a balance of all three parts, for each is a necessary component of our existence and each must be given its due. But paramount over the other two must be the intellect, or reason, for it alone can properly guide our lives. Otherwise we will run wild with impulsive and thoughtless acts that can only reduce whatever chance we might otherwise have for being happy.

The Greeks firmly believed that science, as a quest for truth, can and must serve humanity's highest moral aspirations. They understood that by pointing out the vulnerabilities of the human psyche, we can be saved from the personal and national tragedies we are all too prone to author.

PART III:

FROM ANCIENT GREECE TO THE MODERN WORLD

Chapter 14

GREEK SCIENCE IN ROMAN HANDS

D uring the third century BCE the pendulum of classical history started to swing from the Greek east to the Roman west as Rome began to realize its imperialistic ambitions, conquering the Greek colonies of southern Italy and Sicily and twice defeating the North African kingdom of Carthage, its archrival for the commercial control of the western Mediterranean. In 146 BCE, the Romans obliterated Carthage and, with the destruction of Corinth the same year, proceeded to dominate mainland Greece.

CULTURE WARS

Rome's dramatic military ascendancy brought it face-to-face with Greek civilization and the cultural values and achievements it stood for, values and achievements radically different from those of the Romans themselves. This confrontation also presented the Romans with a psychological challenge they had never before encountered, a challenge to their national self-esteem.

Though the Greeks grudgingly acknowledged Rome's military superiority, they ridiculed what they saw as Rome's cultural inferiority. After all, when fifth-century BCE Athens had been reveling in a golden age of intellectual and artistic splendor, Rome had been a crude and unsophisticated town lacking any semblance of intellectualism or art. When in the fourth century BCE Alexander the Great went on to inspire a new era of Hellenistic creativity, Roman Italy was still a cultural backwater. If Rome was to hold its head up high among the nations of the world and not be humiliated by the very peoples it defeated, it would need to rapidly acquire at least the veneer of cultural respectability.

185

Many leading Romans, like Lucius Aemilius Paullus, favored the whole-sale adoption of Greek culture, went on to learn Greek, studied rhetoric and philosophy abroad, and engaged Greek tutors for their sons. But others, like Cato the Elder, saw such moves as dangerously subversive because the core values of Greece were diametrically opposed to those of Rome.[1] Instead of the material betterment of everyday life in which the Romans believed, the Greeks emphasized the human spirit and its realization through the less tangible pursuit of excellence. Instead of the self-assured application of unlimited force to transform the external world, the Greeks stressed the inner search for self-knowledge and the necessity of moderation. Instead of respecting authority in its own right and giving its holders absolute obedience as did the Romans, the Greeks passionately loved freedom and embodied restless curiosity. Last and most important, to the Greek mind the individual and not the state commanded man's highest allegiance. Appropriately, the most sacred place in Athens was the Parthenon, the temple of Athena, a goddess who represented the exercise of critical intelligence, whereas the most holy place in Rome was the sanctuary of the goddess Vesta, a goddess of the hearth and home who symbolized dutiful submission to parental authority.

What censorious Cato feared at that time was later set in verse by the Augustan poet Horace: "Conquered Greece captivated its untamed conqueror" (*Graecia capta ferum victorem cepit*).[2]

Ultimately, however, the historical response of the Romans was to select the products of Greek civilization that suited their ethnic taste, adapting what they borrowed to serve their cultural needs. The rest they astutely rejected as unworthy of Roman national character. In literature, the Hellenic genres of history, epic poetry, and oratory were put to good use to glorify and fulfill their nationalistic ambitions. In art, the most lavish of Greek architectural styles, the Corinthian, was favored for the spectacular look it gave to monuments, while sculpture was exploited to gratify the egos of the powerful through portraiture or to trumpet their achievements through historical relief. The works of art themselves were largely executed by enslaved Greek prisoners of war or voluntary Hellenic expatriates who labored anonymously to fulfill commissions from their Roman masters. Through this process of cultural assimilation, as we shall see, the Romans would unconsciously perform the vital service of preserving, organizing, and transmitting the heritage of Greece to later ages.

THE ROLE OF SCIENCE

Perhaps the sharpest distinction between the cultural attitudes of the Greeks and Romans can be seen in their contrasting attitudes toward science. First and foremost, the intellectualizing Greeks pursued knowledge for its own sake, delighting in the infinite powers of the mind and the profound truths they could reveal. Only secondarily, as for example in medicine and mechanics, did they consider the practical applications of their discoveries. The down-to-earth Romans, for their part, put utility above all, especially to the degree that such knowledge could help them govern the world and create a secure and prosperous society. One was a race of explorers, like Odysseus; the other, a race of nation builders, like Aeneas (fig. 41). One people was

Figure 41: Augustus Caesar, founder of the Roman Empire, who traced his ancestry to the legendary Aeneas. From Franz von Reber, History of Ancient Art, rev. ed., trans. Joseph Thacher Clarke (New York: Harper & Brothers, 1882).

driven by boundless curiosity; the other, by a need to control and fulfill their global destiny. Repelled by theory and abstraction, the Roman mind gravitated instead toward concreteness and practicality.

If a job required special expertise, the Romans would more often than not hire a Greek to do it—just as Julius Caesar did when he hired an Alexandrian astronomer to reform the archaic Roman calendar. Such behavior did not make the Romans less civilized than the Greeks; it simply meant they had a realistic grasp of their own limitations. What made the no-nonsense Romans strong was that very realism, a dynamically pragmatic trait that set them apart from other ancient cultures.

The Roman poet Virgil, who perfected his craft by emulating Homer, said it best. Here, in a scene from the *Aeneid* set in the underworld, the ghost of Aeneas's father, Anchises, tells his son—and through him the generations of Romans that would follow—not to be overwhelmed by a sense of inferiority when comparing himself to the Greeks. A civilization can only succeed, Virgil says, by identifying and actualizing those cultural traits that are uniquely its own. Referring to the Greeks dismissively, Anchises (i.e., Virgil) declares:

> Let others *make the bronze of statues breathe,*
> *as only* they *can do. Let* them *sculpt out of marble*
> *faces that live. Let* them *weave fancy words*
> *and picture the paths of planets and stars. Make it*
> your *business, Roman, to rule the world with the means*
> that are yours: *enforcing peace, sparing the submissive,*
> *and crushing the arrogant in war.*[3]

A century later, and swelling with civic pride, Sextus Julius Frontinus, the water commissioner of Imperial Rome, would lavish praise on his city's aqueducts (fig. 42). "With so many indispensable structures transporting so much water compare, if you will, the patently useless Pyramids or the equally worthless, though much celebrated, works of the Greeks!"[4] To a Roman engineer like Frontinus, a water-delivery system (however mundane it might appear) that served the public good was a truer hallmark of civilization than the tomb of a dead pharaoh, however monumental, or the marble temple of a remote god (plate 8).

Frontinus's bias aside, even a tough Roman general like Marcus Claudius

Figure 42: The ruins of the third-century BCE Aqua Appia, Rome's earliest aboveground aqueduct. From Thomas H. Dyer, The City of Rome: Its Vicissitudes and Monuments, *2nd rev. ed. (London: George Bell & Sons, 1883).*

Marcellus could recognize and acknowledge intellectual greatness when he saw it. According to the biographer Plutarch, Marcellus "was a warrior by trade, with a powerful body and a brawler's fists, who by temperament was fond of battle, where he displayed both pride and ferocity. But otherwise he was a humane individual who practiced moderation and loved Greek literature and philosophy, so much so that he admired and respected those who excelled at them seeing as he himself had been prevented from engaging in them because of his occupation."[5]

The greatest proof of Marcellus's respect for Greek learning came during the course of the siege of Syracuse in 211 BCE, when the genius Archimedes, who had aided the Syracusan cause, lost his life.

What distressed Marcellus [then the Roman commander-in-chief] the most was the fate that befell Archimedes. As it happened, Archimedes was by himself, preoccupied with a problem he was working out with the help of a diagram. So focused was he on the problem that he was unaware that the Romans had broken in and captured the city. Paying no attention to a soldier who came up and ordered him to follow him to Marcellus' headquarters, Archimedes said he didn't want to go until he had worked out the solution and established the proof. At that, the soldier, flying into a rage, drew his sword and killed him. Others report that the Roman soldier had already drawn his sword and was about to kill him when Archimedes begged and pleaded to let him stay a little longer so he wouldn't have to leave his work behind half-done. But the soldier paid no attention and killed him anyway [plate 9]. According to a third version, Archimedes was in the process of transporting to Marcellus the scientific instruments—the sundials, spheres, and quadrants—he used to measure the size of the sun when some soldiers came upon him and killed him, thinking he had gold in the box. All concur, however, that Marcellus was pained at the news of Archimedes' death, turned away from the murderer as though from a man who bore a curse, and proceeded to find and honor Archimedes' relatives.[6]

With the long passage of time, the grave of the Greek scientist lay abandoned. Ironically, it took another Roman to find it and, out of respect, save it from oblivion. The discoverer was none other than the orator Cicero (106–43 BCE), who, in admiring the mathematics of the Greeks, readily admitted that his own people had failed to honor mathematicians or appreciate the grandeur of their art, limiting it instead to mere mundane measurements and calculations.[7] In the following passage, Cicero (fig. 43) recounts the homage he paid to Archimedes, 137 years after the man's death.

Figure 43: Bust of Cicero. From Francis W. Kelsey, ed., Select Orations and Letters of Cicero, 9th ed. (Boston and Chicago: Allyn and Bacon, 1892).

When I was provincial administrator in Sicily, I myself tracked down his grave, hedged in as it was on all sides by briars and brambles, a burial-place—I would add—that the Syracusans themselves were oblivious to, denying that it even existed. Recalling a few verses supposedly inscribed on his tomb that said it was topped by a sphere with a cylinder, I scanned the whole cemetery (for there are a plethora of graves at the Agrigentine Gate) and noticed a modestly-sized column rising a bit above the surrounding thicket, and on it the outline of a sphere and cylinder. I at once informed the Syracusans (for a delegation had accompanied me) that I thought it was the very thing I was looking for. Slaves were then sent in with sickles to clear the area and, when a pathway was opened up, we approached the pedestal that faced us. Though the last parts of the verses were almost completely worn away, you could still make out the rest of the original epigram.

Thus the noblest city of the Greek world, once renowned for its learning, would have been oblivious to the last resting place of its single most brilliant citizen had not a man from the little Italian town of Arpinum not come and pointed it out.[8]

Note, however, in these proud last lines how Cicero simultaneously praises Archimedes and denigrates the modern Syracusans for their negligence. Cicero asserts that it took a determined Roman—him—to do what no Greek could. ("Ye gods"—to paraphrase the man from Arpinum—"it was right there in front of their eyes if only they had opened them!")

At their best, however, the Romans acknowledged what only the Greeks could teach them. The engineer Vitruvius, for example, in his first-century BCE handbook on architecture, asserted that a proper architect needed a well-rounded humanistic education. "He should have a liberal education and be a man who is an expert draftsman, educated in geometry, familiar with works of history, and a diligent student of philosophy, one who knows music, has some understanding of medicine, is learned in the law, and is conversant with astrology and astronomy."[9]

But in typical Roman fashion, Vitruvius went on in the next breath to explain the pragmatic advantages of just such humanistic training.[10]

THE ENCYCLOPEDISTS

One of the greatest gifts of the Romans was their talent for organization. Without it, they could not have built and administered a three-thousand-mile-wide empire that lasted for more than half a millennium, nor could they have created colossal masterpieces of engineering and urban architecture that endure to this day.

The Romans brought this same talent for organization to bear in comprehending and utilizing the scientific discoveries of the Greeks. Unlike Greek rationalism, whose principal aim was abstract understanding and whose supreme product was philosophy, the Roman aptitude for organization was directed at remaking the real world and its inhabitants. Toward that end, various Roman authors translated and synthesized Greek thought, repackaging it in forthright Latin for domestic consumption and use. Transmuted in this way, the pure science of Greece eventually became the applied science of Rome. Indeed, through this process the Latin language itself was refined from its earliest crude form, as Roman writers were forced to come to grips with the clarity, precision, and literal meaning of the Greek language.

Two notable poets of the Augustan age, Lucretius (ca. 94–55/51 BCE) and Ovid (43 BCE–17 CE), composed panoramic epics in which they took two differing fundamental principles of the universe, first enunciated by Greek philosophers centuries earlier, and illustrated them in compelling depth and detail. Inspired by the atomic theory of Democritus (ca. 460–370 BCE), which had been converted into a guide for life by Epicurus (341–270 BCE), Lucretius in *De Rerum Natura* (On the Nature of Things) majestically discoursed on the themes of matter and space, the movements and shapes of atoms, the human mind and spirit, sensation and sex, the emergence of life and society, and celestial and terrestrial phenomena. Lucretius's scientifically rational analysis of the world was intended to free men and women of their superstitious fears by showing them the non-supernatural basis for all human experience. For Ovid, the defining principle of the universe was transience, a principle first proclaimed by the Greek philosopher Democritus ("The only permanent thing is change."). In his twelve-thousand-verse epic *The Metamorphoses*, Ovid created a vast and variegated tapestry out of the strands of classical myth and legend, seamlessly interweaving over a hundred separate tales of transformation from the origin of the universe to the birth of the Roman Empire.

A third Latin poet, alas anonymous, explored in his *Aetna* the reasons for volcanic eruptions, eschewing mythological explanations and choosing instead those of science. In the midst of his versified treatise, he eloquently described science as one of humanity's most noble and necessary undertakings. He proclaimed:

> *Our duty is not to gaze upon the world's wonders*
> *with the eyes of a dumb brute,*
> *or feed our ponderous flesh*
> *when prostrate on the lowly ground.*
> *Our duty instead is to learn the ultimate truth and the final cause,*
> *holding our minds sacred while lifting our heads to the skies;*
> *to know what principles were present at the earth's inception*
> *(Do they fear extinction, or do the ages march on,*
> *their engine secured by an adamantine chain?);*
> *to know the length of the sun's circuit and the lesser length of the*
>   *moon's*
> *(the latter whose yearly travels take twelve orbits,*
> *while the former's take but one);*
> *to know what celestial bodies follow a set path*
> *and which ones depart with inconstancy from their course;*
> *to know the succession of the constellations and the laws they obey;*
> *why the seasons change (why spring dies*
> *in the prime of its youth as summer draws near;*
> *why summer fades and winter steals up on autumn; and why the cycle*
>   *recurs);*
> *to know the currents of the sea and predict the voyages of the stars.*
> *Our duty then is to not permit the cosmic miracles that exist*
> *to lie disjointed in disarray,*
> *or be buried beneath clutter,*
> *but to arrange them each in its own clearly marked location,*
> *for such can be humanity's divine and joyous delight.*[11]

Two other writers, separated by five centuries, designed model curricula for the training of Roman youth based on the liberal arts first developed by the Greeks. The most learned scholar the Republic ever produced was Marcus Terentius Varro (116–27 BCE). Singled out by Julius Caesar to be the founder and director of Rome's first public library, Varro authored a number of systematic treatises on the development and structure of the Latin language, on agricultural methods, and on ancient (i.e., "ancient" to Varro) Italian customs and beliefs. Regrettably, all of these works are now lost together with *The Disciplines*, a nine-volume dissertation in which Varro, taking his lead from Greek philosophers, delineated the academic subjects a man must master if he wished to consider himself truly educated. These included, we are told, grammar, rhetoric, dialectic (i.e., philosophical argumentation), arithmetic, geometry, astronomy, music, medicine, and architecture. Still surviving intact, however, is *The Marriage of Philology and Mercury*, an elaborate treatise on the same theme by Martianus Capella, a lawyer who lived in Roman North Africa during the fifth century CE and who composed his masterpiece as the world was crumbling around him after the barbarians sacked Rome in 410. This allegorical work, dedicated to Capella's son, describes how virginal Philology, the study of language and literature, married Mercury, the god of profitable pursuits (in other words, it showed how someone can earn a living yet be intellectually well-rounded). At the celestial wedding banquet, seven divine bridesmaids take turns describing the (now) *seven* liberal arts: Grammar, Rhetoric, Dialectic, Arithmetic, Geometry, Astronomy, and Music. (Though Capella patterned his curriculum on Varro's *Disciplines* and allowed Medicine and Architecture to attend the banquet, he didn't give Medicine and Architecture any lines to speak, believing them too mundane.)

High on our list of Roman encyclopedists are Seneca the Younger (4 BCE/1 CE–65 CE) and Pliny the Elder (23/24–79 CE). Seneca was a Stoic moralist and essayist who had the unenviable role of tutoring the maniacal Emperor Nero, an untenable teaching post that eventually cost Seneca his life. Charged by the emperor with conspiracy, Seneca was forced to commit suicide.

As a science writer, Seneca's claim to fame rests on his *Inquiries about Nature*, or *Natural Questions* (Naturales Quaestiones), a wide-ranging discussion in Latin (with moralistic asides) that was based on Greek sources. The work covers lights in the sky (including rainbows, halos, and shooting stars), thunder and lightning, terrestrial waters (including a separate chapter on the Nile), snow and hail, the winds, earthquakes, and comets. Significantly, Seneca anticipated the advance of science. "There will come a time," he wrote, "when those things that are now hidden will be brought to light by diligent research . . . and revealed over the long course of generations. There will come a time when our descendants will marvel at the seemingly obvious things we did not know. . . . Let us therefore be satisfied with what we have found; and let our descendants add to the search for truth."[12]

Unlike Seneca, Pliny the Elder died from a mixture of curiosity and duty, perishing from poisonous volcanic fumes while commanding a rescue mission on the Bay of Naples during the eruption of Mount Vesuvius in 79 CE. According to the testimony of his nephew, he was a man of indefatigable intellectual energy and concentration, who at great risk to his life continued to dictate to his personal secretary the progressive phases of the deadly eruption even as projectiles from the volcano were raining down around him. His masterpiece was his *Natural History*, "perhaps the most important single source extant for the history of ancient civilizations."[13] Drawn from hundreds of ancient sources now lost to us, the 2,500-page work includes some twenty thousand individual topics, investigations, and observations covering such subjects as anthropology, astrology, astronomy, botany, entomology, geography, geology, ichthyology, medicine, mineralogy (including magnetism), ornithology, painting, pharmacology, physiology, psychology, spiritualism, and zoology. As Lynn Thorndike has noted, our word *experiment* comes from the Latin *experimentum*, used by Pliny to refer to the information he acquired from direct personal experience or, in some cases, from efforts to scientifically verify a proposition.[14]

In his preface, Pliny proudly declares that no Roman had ever before attempted so vast a project, nor had any Greek ever tried on his own to deal with so many separate aspects of what the Greeks called "encyclopedic knowledge" ("Preface," 14). Despite Pliny's boast, there had been a Greek who had blazed the trail, the leading Stoic philosopher Posidonius (ca. 135–ca. 51 BCE), who popularized such diverse subjects as anthropology, astronomy, botany, geography, history, hydrology, mathematics, meteorology, seismology, and zoology.

As Classicist Ian Gray Kidd observes: "Posidonius' position in intellectual history is remarkable not for the scattered riches of a polymath and savant, but for an audacious aetiological attempt to survey and explain the complete field of the human intellect and the universe in which it finds itself an organic part, through analysis of detail and the synthesis of the whole, in the conviction that all knowledge is interrelated."[15]

Paradoxically, Pliny's magnum opus is the consummate proof of Rome's greatest strength in the field of science and simultaneously a demonstration of its greatest weakness. The Romans were eager to collect, organize, and apply the discoveries of the Greeks, but they were not prone, despite Seneca's pious hope, to follow the path to further discovery. As Seneca himself sadly noted at the conclusion of his work: "Nowadays people devote so little attention to the topics that the ancients probed that even the few things they once discovered are being forgotten. But by Hercules, if we pursued that endeavor with all our might, if the younger generation took it seriously and hearkened to what the older generation taught, we'd get to the bottom of things and find where the truth lies instead of nonchalantly skimming the surface."[16]

Pliny agreed, asserting that Rome's materialistic creed had killed the spirit of scientific inquiry that once flourished in Greece. He wrote: "Most of those Greeks who made these discoveries did so with no other reward in mind but the benefits they could bestow upon posterity. Today, however, the passage of time has weakened man's character, though not his commercial ambition. Now that every sea and shore lies open to exploitation, multitudes go forth on voyages—but for the sake of lucre, not learning."[17]

Within centuries, however, a dark age would descend upon Rome and the empire it ruled. Still, as we will see, it was the Roman systemization of Greek scientific knowledge that enabled it to survive and, despite new challenges, generate a revival of scientific learning.

Chapter 15

THE TRANSMISSION OF GREEK SCIENCE TO LATER AGES

Over the course of six centuries, from the subjugation of Greece by the Romans in 146 BCE to the fall of Roman Italy to the barbarians in 476 CE, Rome amassed a fortune in Greek learning that became a major portion of its legacy to the later world.

STOREHOUSES OF KNOWLEDGE

During the second and first centuries BCE a number of distinguished bilingual Romans—including Scipio Aemilianus, Sulla, Lucullus, and Cicero—began to avidly collect Greek and Latin books, motivated in good measure by the enhanced social prestige such private collections would confer upon them. Inspired by the earlier example of Hellenistic monarchs who had founded great public libraries in Alexandria and Pergamum, Julius Caesar contemplated founding a similar institution at Rome under Varro's eminent leadership. The dictator's assassination cut these plans short, but they were revived a few years later by the consul Asinius Pollio, who erected Rome's first public library in 39 BCE from the spoils of war. Its construction was followed by the rise of libraries sponsored by Augustus Caesar (27 BCE–14 CE) and his imperial successors, including Trajan (98–117 CE), who incorporated into the design of his magnificent new forum a library with two wings—one for Greek books and another of equal size for Latin books. By the time of Constantine the Great (312–337 CE), Rome boasted twenty-eight public libraries housing Greek and Latin literary and scientific works.

Other public libraries could be found scattered throughout Italy and Rome's empire, often thanks to the generosity of civic-minded citizens. As for private libraries, the most extraordinary (because of its rediscovery) is the library in the so-called Villa of the Papyri at Herculaneum, buried by the eruption of Mount Vesuvius in 79 CE and disinterred in the 1750s. Most of the papyrus scrolls contain works, including scientific writings on natural phenomena, by the Epicurean Greek philosopher Philodemus, who immigrated to Italy in 80 BCE. (The Epicureans regarded the pursuit of life's most lasting pleasures as their chief aim.) The villa's scrolls were deeply charred by volcanic heat and therefore extremely fragile. With the help of an ingenious mechanical device, some were unrolled a millimeter at a time. To distinguish the inked letters from their blackened background, modern investigators are using a multispectral imaging technique developed in the 1990s that employs infrared and ultraviolet filters. In the future, CT scans may succeed in penetrating the inscribed inner layers of scrolls that are too brittle to even attempt to open. As for the villa itself, its layout has inspired the architectural design of the Getty Villa in Malibu, California, the former home of the Getty Museum.

The printing press would not be invented until the fifteenth century, so all ancient books were handwritten and therefore were far fewer and less widely dispersed than are books today. While the concentration of books in libraries undoubtedly advanced the cause of learning, this accumulated knowledge was more susceptible to wholesale destruction by fire, whether accidental or deliberate.

Another reason such knowledge was vulnerable was the fact that its mere existence did not guarantee that it would be used or held in high esteem, especially in the days of Roman civilization's decline. When those twenty-eight libraries still stood, the writer Ammianus Marcellinus (ca. 330–395 CE), Rome's last great Latin historian, saw in the degeneration of his culture a parody of show business:

> Those few institutions of learning which were once celebrated for their serious cultivation of intellectual pursuits are now filled with inane and idle diversions that echo with voices warbling to the strains of lyres and flutes. Today, instead of a philosopher you'll find a music teacher; instead of an orator, an acting coach. And as for the libraries, they're sealed perpetually like tombs. . . . The Romans have sunk so low that, not long ago, when the

government anticipated a famine and foreigners were told to get out of the city, people pursuing the liberal arts were peremptorily expelled, while the actresses and their entourages were allowed to stay. Three thousand strippers weren't even brought in for questioning. In fact, along with the chorus-girls and choreographers, they were declared exempt.[1]

BARBARIANS AT THE GATES

As Ammianus Marcellinus's remarks imply, the eventual fall of Rome was due as much to an internal dissolution of national character as it was to the external assaults of barbarians. But assaults there were.

By the end of the third century CE, following Diocletian's desperate partition of Rome's crumbling Mediterranean empire into two administrative halves, the northern and strategically located city of Milan (ancient Mediolanum) replaced Rome as its western capital. In the year 378, during Ammianus Marcellinus's lifetime, the once invincible Roman army was defeated by the Visigoths in eastern Europe. In 387, "impregnable" Rome was sacked by the Gauls. By 403, the Roman emperor of the West abandoned Milan for the more topographically defensible site of Ravenna near Venice. By 406, barbarian tribes overran France. The next year, Roman troops pulled out of Britain. In 410, the Visigoths sacked Rome, leading St. Jerome to declare: "The city that had conquered the entire world was itself now conquered."[2] In 450, Milan was destroyed by the Huns. Five years later, the Vandals pillaged Rome, thereby earning a place in infamy as wanton destroyers. In 476, the last Roman emperor of the West, a mere boy named, ironically, Romulus (after the mythic founder of Rome) Augustulus ("Little Augustus") abdicated his throne to a barbarian chief at Ravenna. In 546, a decimated Rome was ravaged for the fourth time, this time by the Ostrogoths. Despite a brief cultural reprieve granted by the enlightened Ostrogothic leader Theodoric, and the flickering hopes it ignited in the heart of the Roman philosopher and translator Boethius, night had finally fallen on Imperial Rome.

THE TASK OF PRESERVATION

We cannot say what happened to all the ancient books during these last chaotic centuries when sheer survival must have mattered far more to people than books ever could. Rome's urban population, once numbered at a million in the days of Augustus and Trajan, had shrunk to a mere thirty thousand. The odds against the survival of books were surely high: those that escaped the barbarian's torch, the harshness of weather, and the insatiable appetite of worms could still be tossed on a humble hearth to cook supper or keep a family warm. And it is doubtful that devotion to theoretical science would have held much fascination for an increasingly illiterate population of a dark age, whose Hobbesian lives were "poore, nasty, brutish, and short."[3] Yet within those neglected books, like insects caught in glowing amber, was the wisdom of a lost world.

The vacuum created by the implosion of the Roman Empire was soon filled by the Christian Church. First persecuted by the Roman state as subversive but later befriended by Constantine for the support it could give, Christianity with its hierarchical administrative structure and promise of a better life gave people a cohesive system they could once again believe in, one in which Latin, the tongue of the former pagan empire, endured as the language of prayer, and Roman law lived on in the canon law of the Church. Though some Christian leaders regarded the Roman Empire as inherently evil, others acknowledged its accomplishments and the civilizing influence its literature and philosophy had on the pagan mind, and the pressing need to salvage and rescue its artifacts. In this endeavor, medieval monasteries would lead the way—preserving and painstakingly copying ancient manuscripts, in part to honor the memory of the past, but also as a spiritual exercise.

Yet just as the Romans had selectively borrowed from the legacy of the Greeks, so did Christian copyists and scholars single out those classical works that most appealed to their own cultural and religious sensibilities. As Gilbert Highet relates in *The Classical Tradition*:

> Certain authors were taught to advanced students and commented on by advanced teachers. But many, many other authors were lost, in part or wholly, forever. Pagan authors were much less likely to survive than Christian authors. Informative authors were much more likely to survive than emotional and individual authors. Thus we have still the works of many

unimportant geographers and encyclopaedists, but hardly any lyrical and dramatic poetry—although in the Greco-Roman world at its height there was far more emphasis on pure poetry than on predigested information. Moral critics were likely to survive, but immoralists not. . . . Also, the scholars of the Dark Ages were more inclined to read and copy authors nearer to them in time. . . . Therefore they devoted a great proportion of their time and energy to authors who are comparatively unimportant but who lived near their own day.[4]

And because Latin was the language of the Church of Rome, works in the original Greek—assuming they could be read at all—were given short shrift. In Highet's words: "Greek remained almost a closed field. Again and again one finds that the medieval copyist who writes Latin correctly and beautifully breaks down when he comes to Greek: he will copy a string of gibberish, or add a plaintive note saying 'because this was in Greek, it was unreadable' [for] the knowledge of Greek in western Europe died out almost completely in the Dark Ages. . . . Aristotle was read, not in the language he wrote, but in Latin translations."[5]

The survival of scientific literature faced particular challenges. Christianity focused on scripture rather than the natural world—the preeminent province of the scientist. It emphasized spiritual faith over critical reason and the acceptance of higher authority over curiosity and intellectual independence. In short, despite its role as a savior of precious documents, the ascendancy of the Church marked an ideological retreat from both the scientific attitude of the Greeks and the very subject matter of their scientific investigations. St. Augustine (354–430 CE) perhaps put it best. Though elsewhere acknowledging his intellectual debt to classical philosophy, Augustine wrote:

When we are asked then what we should believe when it comes to religion, we need have no recourse to examining the nature of the universe as did the so-called natural philosophers (*physici*) of Greece, nor should we fear being ignorant as Christians of the power and number of the elements; the motion, order, and eclipses of celestial bodies; the configuration of the heavens; the varieties and nature of animals, plants, rocks, fountains, rivers, mountains; the measurement of space and time; the warning signs of bad weather; and all the hundreds upon hundreds of other things those Greeks either discovered or thought they had discovered, since they themselves— despite their immense intellect, blazing zeal, and ample leisure—did not uncover everything, whether their investigations were based on a priori speculation or empirical evidence, for they prided themselves on discoveries that reflected conjecture more than real knowledge.

It is enough for a Christian to believe that the origin of all things created, whether in heaven or on earth, whether visible or invisible, is nothing other than the goodness of the Creator, who is the one true God, and that no nature exists apart from Him and His will.[6]

Additionally, as Classicist Marshall Clagett has observed, the "growth and triumph of the Church siphoned off men who [otherwise] might well have pursued natural philosophy or science."[7]

LIGHT FROM THE EAST

Simultaneous with these developments in Christian Europe, other events were taking place in the Middle East, where the religion of Islam would arise in the seventh century CE. Because of the eastward reach of Alexander the Great's conquests in the fourth century BCE, Hellenistic culture had penetrated the very lands where Islam came to flourish. As a result, Greek works, including texts on natural philosophy, mathematics, astronomy, and medicine, circulated and were translated, first into Syriac and then into Arabic, especially in Baghdad under the enlightened patronage of the Abbasid court. Indeed, during the eighth, ninth, and tenth centuries CE, Arabic translations and commentaries were made of nearly all the works of Greek science with which we are familiar today. Perhaps the greatest figure of all in this intellectual era was the Persian philosopher and scientist Ibn Sina (980–1037 CE), better known in the West as Avicenna.

Because of the Moorish conquest and occupation of Spain beginning in the eighth century, Muslim intellectuals brought these works with them to Europe, where a golden age of Islamic civilization came to flourish on Iberian soil. By the early ninth century, Muslims had also become politically and culturally dominant in Sicily. Many of their scholarly works that stemmed from the classical tradition were in turn translated by Christian thinkers into Latin, and came to play an influential role in shaping scientific thought in the newly founded universities of western Europe.

As Thomas Goldstein states in *Dawn of Modern Science*: "The resurgent image of the classical world—and, with it, the first stirrings of historical consciousness—came to the European mind very largely in the shape of science. . . . If Europe managed to lift itself from the status of a half-barbaric society to a center of pulsating creative culture, those crude translations of scientific texts played a vital role: they helped in piecing together the shattered image of the classical world and reweaving the sense of cultural continuity."[8]

In this process of intellectual transmission, by far the greatest Islamic proponent of Aristotle's writings was the Spanish scholar Averroes, or Ibn Rushd (1126–1198 CE), who said that the great philosopher "comprehended the whole truth—by which I mean that quantity which human nature, insofar as it is human, is capable of grasping."[9] Indeed, to settle an intellectual argument at a medieval university, it was enough to say *Magister dixit* (The Master [i.e., Aristotle] said it).

In order for the free and independent spirit of Greek science to fully reinhabit the Western mind, a revolutionary environment was needed that would validate and revitalize the fundamental principles the Greeks had believed in. It would take almost a millennium before that environment would reemerge. But before it did, other remarkable developments were already taking place in the eastern Mediterranean.

TREASURES FROM BYZANTIUM

Before Rome's European empire collapsed, the Roman Empire had been officially divided into two geographic halves, with Constantinople (formerly Byzantium) functioning as its eastern administrative center. When the Western Empire fell into barbarian hands, the Eastern, or Byzantine, Empire endured and lasted for another thousand years until its conquest by the Ottoman Turks in 1453. Though Constantinople was threatened time and again by barbarians, their forces never breached the city's walls, leaving it secure. Most notable among Constantinople's rulers were its namesake, Constantine, and the emperor Justinian (527–565 CE) and his wife, Theodora, who presided over a brilliant era of intellectual and artistic creativity.

Unlike the West, where the predominance of Latin during the dark ages effectively alienated scholars from writings in Greek, in the Byzantine Empire Greek continued to be the dominant language in the same way that it had served as the lingua franca of the eastern Mediterranean ever since the Hellenistic age. In addition to linguistic continuity, the conservation and cultivation of Greek manuscripts and their scholarly study and elaboration through commentaries was deliberately encouraged by an enlightened Byzantine court, especially those subject areas in which Greek learning could benefit society.

Although the great library at Alexandria had been lost to war around 270 CE, the city of Constantinople possessed three impressive libraries. One of these held strictly theological works, but the other two were secular in nature: a library at the palace that served the needs of the imperial family and their courtiers, and a library attached to an imperially sponsored "university" where language, literature, and law were taught. Among the bibliographical treasures of Constantinople were the works of Plato, Aristotle, Euclid, Archimedes, Hero of Alexandria, and Ptolemy—all in the original Greek. Meanwhile in the city's schools, younger students were instructed in the traditional seven liberal arts, including mathematics, music, and astronomy.

At the beginning of the thirteenth century, in the days of the Fourth Crusade, the legacy of learning that had been guarded for centuries by the city of Constantinople was ruthlessly savaged, not by an army of Muslims or barbarians, but by an army of Christians.

According to Reviel Netz and William Noel:

Nicetas Choniates, the brother of the Archbishop of Athens, witnessed the greatest calamity that ever befell the world of learning. In April 1204, Christian soldiers on a mission to liberate Jerusalem stopped short of their goal and sacked Constantinople, the richest city in Europe. Nicetas gave an eyewitness account of the carnage. The sumptuous treasure of the great church of Hagia Sophia (Holy Wisdom) was broken into bits and distributed among the soldiers. Mules were led to the very sanctuary of the church to carry the loot away. A harlot, a worker of incantations and poisonings, sat in the seat of the Patriarch and danced and sang an obscene song. The soldiers captured and raped the nuns who were consecrated to God. "Oh, immortal God," cried Nicetas, "how great were the afflictions of the men." The obscene realities of medieval warfare crashed upon Constantinople, and the hub of a great empire was shattered.[10]

THE *ARCHIMEDES CODEX*

Among the precious objects that survived the burning and pillaging of the city were books containing the thoughts of Greece's greatest scientists. Among these were three codices, or bound manuscripts, of Archimedes's works written in the original Greek. Today, though many copies of Archimedes's works exist in Latin translation, only one parchment codex in the original Greek endures, copied by a Byzantine scribe around the year 900 CE from an even older text. In the year 1229, in need of a new prayer book, a priest erased the words of Archimedes from the pages of the manuscript by scraping and washing off the antique ink, cutting the ninety pages in two, and, before rebinding them, writing his own prayers upon their surface. In doing this, he had created what bibliophiles call a *palimpsest*, a recycled manuscript whose original words have been deliberately obliterated and overwritten (plate 10). Carried from Constantinople to Jordan and then back again to Constantinople, the *Archimedes Codex*, or *Palimpsest* (as it has come to be called), was rediscovered at the dawn of the twentieth century by a Danish scholar, Johan Ludwig Heiberg, who recognized its significance and tried, with the help of black-and-white photography and the limited aid of a magnifying glass, to read the "invisible ink" beneath the medieval prayers. By the early 1920s the codex had passed into the hands of a private collector in Paris, where it remained, stained and blackened by mold, until it was sent to Christie's in New York in 1998 and put up for auction. Sold to an anonymous bidder for $2,000,000, the codex was subsequently entrusted by him to the Walters Art Gallery in Baltimore, where it has been under study for over a decade. There it has been subjected to ultraviolet, infrared, visible, and raking light, as well as X-rays, and digitized to permit scholars to better decipher its faint and elusive writing and thus read Archimedes's mind.[11]

The codex contains three critically important works by the master mathematician and scientist: the only version that exists in the original Greek of his most famous treatise, *On Floating Bodies*, born of the hydrostatic discovery that led him to exclaim "Eureka!," and two treatises modern scholars had never before seen. The first is his *Method of Mechanical Theorems*, which describes his "secret" method of solving seemingly insoluble problems. This method involves using a mathematical analysis of the shapes of physical objects to determine their properties, including their centers of gravity, and using geometry to approximate an understanding of infinity. In this pursuit, Archimedes anticipated the discovery of calculus by Leibniz and Newton some two thousand years later.

The second "lost" treatise is titled *The Stomachion* because of the stomach-churning challenge it presents: to take a square composed of fourteen irregularly shaped pieces and determine, by sheer geometry, how many different arrangements of the same pieces can yield other squares (the modern mathematical answer is 17,152!). The techniques Archimedes would develop anticipated by more than two millennia the modern mathematical sciences of combinatorics and probability, which investigate the possible and likely combinations that can theoretically occur for both objects and events. Significantly in this latter work, Archimedes challenged himself in typically Greek fashion to wrestle with an immensely difficult problem whose solution offered no practical benefit. The very exercise the puzzle would give his mind and the joy of mental triumph he would experience at its completion were reward enough.

Most of Constantinople's surviving manuscripts, however, did not follow the same journey as did the *Archimedes Codex* but took a more direct route to Europe and with greater immediate effect.

East Meets West

Shortly before Christian Constantinople fell to a Turkish army in 1453, many of the city's leading scholars fled to Italy for refuge, carrying with them a precious cargo of original Greek manuscripts that the West had heard of but never seen. Arriving at the time of the Renaissance, these works would add fuel to the philosophical speculations and scientific imaginations of Italy's best minds. Before this, many scholarly Italians, eager for new knowledge, had already begun to take lessons in classical Greek, and those wealthy patrons of the arts, the Medicis of Florence, had even dispatched agents to Greece to search for new manuscripts with orders to buy them "at any price whatever." The arrival of the new Byzantine émigrés and their literary luggage could not have come at a more fortuitous time for the future of scholarship and science.

At the same time, potential new European sources of knowledge were being searched for with uncommon zeal. "The great book-finder Poggio Bracciolini (1380–1459) describes how he would talk his way into monasteries, ask to see the library, and find manuscripts covered with dust and debris, lying in leaky rat-ridden attics: with touching emotion he speaks of them looking up at him for help."[12]

A HUMANISTIC LAMP RELIT

Significantly, the Renaissance proved to be a fertile field for the seeds of Greek thought to fall on, precisely because it was energized by the same dynamic spirit of humanistic self-confidence that had animated the ancient Greeks. While classical civilization, both Greek and Roman, had viewed human nature as something to be proud of because of its creative potential, the Middle Ages had conversely regarded human nature as something to be ashamed of because it was prone to sin. The Renaissance, for its part, asserted once again that human nature was a thing to take pride in, inasmuch as man could, by freely using the divine gift of reason with which he had been endowed, celebrate his Creator and himself.

In this era of multitalented, multifaceted "Renaissance men," surely the most versatile was Leonardo da Vinci (1452–1519), a man whose every achievement—in sculpture, in painting, and in engineering—derived from his knowledge of science: whether it be the anatomical proportions of a statue, the optical perspective of a landscape, the chemical composition of metals and pigments, or the mechanics of a bird in flight. He possessed the keen vision, the sense of wonder, the driving curiosity, and the questioning spirit that were the hallmarks of Hellenic science at its best. And, like his Renaissance compatriots, he had, as his personal notebooks reveal (fig. 44), drunk deeply from the well of Greek learning.[13]

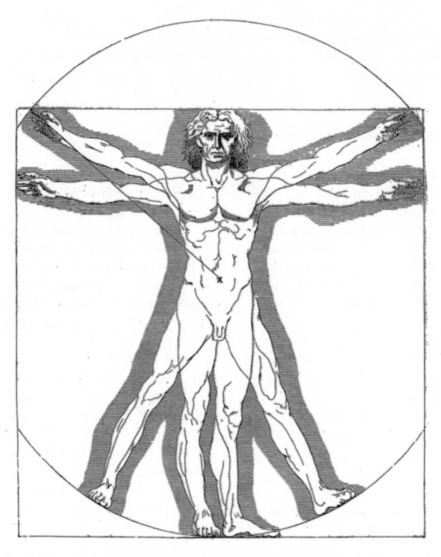

Figure 44: Leonardo da Vinci's Renaissance diagram of man's "geometrical" anatomy, inspired by Vitruvius's description of Polycleitus's canon. From George Redford, A Manual of Sculpture *(London: Sampson Low, Marston, Searle, and Rivington, 1882).*

PART IV:

❧❧ ❧❧ ❧❧ ❧❧ ❧❧ ❧❧ ❧❧ ❧❧

ANCIENT SCIENCE BEYOND THE MEDITERRANEAN

Chapter 16

SKY WATCHERS OF CENTRAL AMERICA

The Mediterranean was the fertile bed upon which the earliest seeds of scientific imagination were sown, first in Egypt and Mesopotamia and later and most energetically in Greece. But beyond the lands of the Mediterranean there were other cultures—in the Old World and the New—that in antiquity strove, each in its own way and each in its own unique environment, to explore the mysteries of the universe.

RUINS IN THE JUNGLE

In the moist and hot jungle of southern Mexico and Guatemala, flesh pulses only to rot. In the dark jade-green tangle of steaming lushness, Death clings to the wet earth, waiting, chorused by the screech of monkeys and the cicada's buzz. Decay is all about, mingled with life. Here only stone does not die, and the stars of the night sky, remote and free. Between them is humanity, caught in time and dissolution. All around, in the leafy rain forest, lurk the hideous masks of the nightmare gods. . . .

In jungle clearings the sacred cities of the Maya rose in profusion, rich in plume-bearing priests and painted shrines, proud chieftains and plans for glory. And then, mysteriously, the cities were abandoned. Slowly, inexorably, the jungle claimed its own, shrouding the deserted temples in impenetrable foliage, choking with thick vines the stone-carved icons of nobles and gods.[1]

Centuries later, when Spanish conquistadors lusting for gold came to the Americas, they heard tales of the lost cities, the very names of which had been forgotten by then. But when no gold was found, their interest waned. In their train came Christian missionaries, eager to convert the natives. But when the natives showed them scrolls with strange symbols salvaged from that lost world, the priests promptly denounced them as works of the devil and set them ablaze. In the smoke of bonfires, the collective legacy of the Mayan sages—except for just four books—vanished for all time.

In the mid-nineteenth century, an American adventurer named John Lloyd Stephens reclaimed the world of the Maya for science, aided in his travels by his English friend Frederick Catherwood, whose evocative drawings and paintings of the ruins inspired other explorers to follow in their footsteps (figs. 45 and 46). Though almost all the painted-bark scrolls had been incinerated, inscriptions in the same hieroglyphic script survived, carved in stone on temple walls and commemorative stelae. Over the course of a century and a half and more, thanks to continuing efforts at decipherment, we have come to appreciate the mind of the ancient Mayan people and its creative brilliance, and have learned that its classic period stretched from about 200 to 900 CE, paralleling the last centuries of the Greco-Roman era and the early Middle Ages of European civilization.

Figure 45: Mayan ruins strangled by roots and vines. From John Lloyd Stephens, Incidents of Travel in Central America, Chiapas, and Yucatan *(New York: Harper, 1841).*

Figure 46: View of the Mayan ruins of Uxmal enclosed by the jungle. From Frederick Catherwood, Views of Ancient Monuments in Central America, Chiapas, and Yucatan *(New York: Bartlett and Welford, 1844).*

The omnipresence of dates in their inscriptions reveals that the Maya were obsessed with time, forever striving to endow their mortal existence with permanence and cosmic significance by associating their earthly deeds with the eternity of the universe. Their complex calendrical system is perhaps their greatest intellectual triumph and the radiant nexus of all their other achievements. The ancient Greeks, for their part, had nothing like it, but then the ancient Greeks chose not to be passively governed by time; instead, they chose to become like gods by defying the limits of mortality.

KEEPING TIME

The Mayan calendar represents the mathematical integration of two separate timekeeping systems: a secular, or civil, calendar of approximately 365 days and a sacred, or religious, calendar of 260 days.

Like a large gear with 365 teeth or cogs, the secular calendar is composed of eighteen months of twenty days followed by a single period of five "unlucky" days. Together, the 365 days reflect the approximate length of the solar year.

The religious calendar, for its part, is like a smaller 260-tooth gear with an even smaller 260-tooth gear rotating inside it. The first of the two small gears consists of thirteen numbers that are repeated twenty times. The second small gear consists of twenty "day-names" that are repeated thirteen times. The two intermeshed gears turn until they meet at their original starting point (after twenty repetitions of the thirteen-number sequence, or thirteen repetitions of the twenty-day sequence). Together, the 260 combinations express dates for ritual observances and holidays.

Making matters more complex, the 360-tooth secular gear is intermeshed with the outer 260-tooth sacred gear. If allowed to run concurrently, all three rotating gears would eventually meet at their original starting point after fifty-two turns of the 360-tooth gear or after seventy-three turns of the 260-tooth gears ($365 \times 52 = 18,980 = 260 \times 73$). Such a complete cycle of 18,980 days is known as the Calendar Round. A person's luck, good or bad, depended upon which sacred gear tooth touched a particular secular day.

Making matters even *more* complex, the Maya also dated events by their distance from a historical starting point, comparable to our use of "CE" (of the Common Era) or "AD" (in the year of Our Lord), both of which begin with the traditional date of Jesus's birth. For Mayan civilization, this key date was 3113 BCE (to be precise, August 11, 3113 BCE), which they regarded as the date the world was created. This period of 5,125 years was referred to as the Long Count. At its conclusion, the world would be destroyed and then created over again. August 11 was chosen because it was a day when the sun passed directly overhead in the latitude where the southern Maya lived.

Because the year 2012 will mark the end of the Mayan Long Count begun in 3113 BCE, many today, already preconditioned by their Christian beliefs to expect an "End of Days," have feared that our world will end in 2012. If you are reading this chapter in the year 2013, you will know that those fears were misguided. If you are *not* reading this chapter in 2013 or after, the Maya might just have been right!

Coincidentally, perhaps, the date 3113 BCE is approximately the time when the first civilizations were born in the Near East. Curiously enough, this date also roughly accords with the traditional Jewish date for the world's creation, 3761 BCE, a date derived from the generations listed in the Old Testament and still used today by observant Jews in marking their holidays.

Since the Maya believed the world had been destroyed and re-created numerous times, there was more than one Long Count. When the end of one Long Count was reached, the Maya odometer "rolled over" and started again at zero. And, because there was more than one Long Count, the Long Counts themselves were numbered. Indeed, actual Mayan dates show that the Maya counted back tens of millions of years into the past and could conceivably have counted tens of millions of years into the future.

Unlike our own Western calendar, every date in the Mayan calendar had secular and sacred coordinates, as well as numbers extended to five place positions to indicate its location in the Long Count. Thus, the ancient date August 11, 3113 BCE, would in Mayan terms have been expressed as 9.16.0.0.0 (its Long Count position) followed by 2 (its sacred day number) followed by Ahau (its sacred month name) followed by 13 (its secular day number) followed by Tzec (its secular month name). The closest analogy would be if an Orthodox Jew today were to simultaneously date every event by both the Western calendar and his own religious calendar, referring to Friday, January 1, 2010, for example, as January 1, 2010, 15 (the day of the equivalent Jewish month) Tevet (the name of that month), 5770 (the number of the Jewish year).

As we can see, the Mayan timekeeping system both required and inspired mathematical accuracy and precision. But since the timekeeping system itself was keyed to the workings of the cosmos, it also required and inspired an accurate and precise knowledge of astronomy. It is for this reason that the Maya became the greatest sun watchers, planet trackers, and stargazers of ancient America.

ASTRONOMERS OF YUCATÁN

Apart from the all-powerful sun, the celestial body that most fascinated the Maya was the planet Venus, the third-brightest body in the heavens, the first to appear after sunset as the Evening Star and the first to herald the sunrise as the Morning Star. Mayan astronomers kept elaborate tables of its heavenly movements (as we see in the surviving "Dresden" Codex) and, without the aid of a telescope, accurately calculated the synodic period of Venus (the time it takes the orbiting planet to return to its same relative position in the sky). Their precise measurement of the length of the solar year (the exact time it takes for Earth to completely circle the sun) was 365.2425 days, only .0003 off from modern astronomy's reckoning of 365.2422.

From the tops of pyramids and towers, Mayan priests scanned the heavens. Indeed, in addition to offering a dramatic focus for religious cere-monies, the height of their public buildings is explained by the need of their priest-astronomers to have a clear view of the horizon unobstructed by tree-tops to perform their sacred tasks.

The Mayas' most famous astronomical observatory is located at Chichén Itzá, the largest ancient city that has been unearthed in the Yucatán. Named El Caracol (Spanish for "conch shell") because of the spiral staircase inside its tower, the observatory supports a turret that rises eighty feet above ground (plate 11). From the turret's windows, astronomers could take sightings of the sun at equinox, of constellations like the Pleiades, and of Venus, a planet they identified with the feathered serpent-god Kukulcán, the bringer of learning and the arts. It is to this god that the largest temple at Chichén Itzá, El Castillo (the Castle), was dedicated. This ninety-foot-tall temple consists of a stepped pyramid with four sides facing north, east, south, and west; a broad staircase on each side; and a sanctuary at the very top. The steps total 365, the number of days in the solar year, while each of the pyramid's four faces bears fifty-two stone panels, the number of years in the 18,980-day Calendar Round. Twice each year at the time of the spring and autumnal equinoxes, the setting sun casts a long shadow along the edge of the north stairway. Because of the angles of the steps, the shadow, as the sun slowly descends, becomes wavy like the body of an undulating serpent, until it dramatically joins itself to the sculpted snake's head of Kukulcán that rests at the pyramid's base.

Unlike the astronomy of the Greeks, the sky watching of the Maya, profound as it was, was a function of communal religious need rather than of individual intellectual curiosity. And that is one of the most crucial differences that set the two cultures apart.

THE PLACE WHERE GODS WERE BORN

Centuries earlier, solar orientation had been used to design an entire metropolis, magnificent in its size and proportions. In the shadow of 16,400-foot volcanoes, the ancient city was dramatically poised atop a 7,550-foot-high plateau overlooking a fertile valley some twenty-five miles north of Mexico City.

The city lay deserted in the days of the Aztecs, who, awed by its monumentality and convinced it must have been built by supernatural giants, named it Teotihuacán, "the place where gods are born." Today, we know the city was founded not by gods but by visionary men in the second century BCE and that it flourished between the second and fifth centuries CE until it was abandoned a century later due to internal unrest, warfare, ecological decline, or some combination of the three. During its heyday, Teotihuacán functioned as a holy city and mecca for pilgrims, boasting a permanent population of between one hundred thousand and two hundred thousand residents. The largest and most impressive city of pre-Columbian America, it was America's first true city.

Covering some twelve square miles, Teotihuacán was laid out on a grid pattern. At its heart was a rectangular two-square-mile ceremonial center, with a long axis that stretched from north to south. On either side of a broad thoroughfare stood temples and shrines. Believing the shrines were tombs, the Spanish conquerors later called the deserted avenue Calle de los Muertos, the Street of the Dead (plate 12).

The largest temple of all was the so-called Pyramid of the Sun, though we do not know with certainty to what deity it was dedicated. Dwarfing the later pyramids of the Maya, it rose in four stepped terraces to a total height of over two hundred feet and occupied a base measuring 730 × 740 feet— half the height but with nearly the same base dimensions as Egypt's largest pyramid, the mid-third millennium BCE Great Pyramid of Cheops. As with Cheops's pyramid, no wheeled vehicles were available to assist in hauling the millions of tons of building materials required for its construction.

The temple's design incorporates optical refinements. To create a greater impression of massiveness in the eye of the beholder, the gradient, or slope, of the four sides actually grows as the structure rises, while the width of its ascending stairs increases to defy perspective and create an illusion from ground level that the sides of the staircases are uniformly parallel all the way to the top. In terms of its orientation, the temple faces in a generally western direction toward the setting sun.

At the northern end of the Street of the Dead stands the smaller, but still massive, Pyramid of the Moon, which rises in four stepped terraces to a total height of 140 feet. But because of the higher elevation of its site, its summit appears equal to that of the Temple of the Sun. The two, however, were built in successive centuries: the Pyramid of the Sun first, and the Temple of the Moon, together with the Street of the Dead, about a century later.

Curiously, the Street of the Dead does not run exactly north to south but is skewed about fifteen and a half degrees so that it runs slightly northeast to southwest. The city's east-west axis is similarly skewed, suggesting that the city was not laid out with a purely solar orientation, but that its founders had some other factor in mind, perhaps one based on sacred landmarks in the distance. Another theory is that the Temples of the Sun and Moon faced toward points on the horizon where during those centuries the Pleiades rose in the east and set in the west, for the Pleiades were the seven-starred constellation whose rising at dawn announced the special day at that latitude when the sun would be directly overhead at noon.

Indeed, the Pleiades may have played a similar role in guiding the orientation of Athens' Parthenon and other Greek temples that traditionally faced in an easterly direction. Whatever the superficial similarities in orientation between the temples of ancient Greece and Central America, the humanistic Greeks distinguished themselves by erecting temples whose sculptures celebrated the heroism of man rather than merely the power of the gods.

Whatever the explanation for Teotihuacán's orientation, excavations at the city's ceremonial center and its environs have gone on for almost a century. But the urban area is so vast, and exploration so expensive and time-consuming that, despite continuing study, perhaps as much as 90 percent of Teotihuacán still lies buried beneath the earth, hiding its age-old secrets. As the whiteness of moonlight illumines the deserted streets, the myriad ghosts of Teotihuacán silently keep their watch.

Chapter 17

SECRETS OF STONEHENGE

Chaos of ruins! Who shall trace the void,
O'er the dim fragments cast a lunar light,
And say, "Here was, or is, where all is doubly night"?
　　　　　　　—Byron, *Childe Harold's Pilgrimage*, Canto IV

With these words Lord Byron addressed the mystery of Stonehenge, one of the oldest monuments in the world and surely one of the most puzzling. Situated on the Salisbury Plain in southern England, its massive stones stand in solitary formation against the sky, looking ever so much like an interrupted domino game played on a table eons ago by the children of giants.

Unlike so many other archaeological wonders that have had to be unearthed, Stonehenge has always been there, taunting both the bused-in tourist and the academic expert. "What am I? Why was I built? Who built me, and how?" asks this English Sphinx.

Throughout history, theories have abounded ever since the days of the Greeks and Romans. In the first century BCE, the Greek historian Diodorus Siculus described an island in the northern sea where priests of Apollo, the sun god, worshipped at a temple "spherical in shape."[1] Diodorus's Roman contemporary Julius Caesar described a priestly caste of Britons (Diodorus's priests?) who spent their days discussing the movements of the stars and the sizes of Earth and the universe.[2] Here Caesar was referring to the Celtic priests of his day called Druids, who, besides reverencing mistletoe, were known to practice human sacrifice. It would not be long before popular imagination would connect them with Stonehenge and make Stonehenge the site of their bloody rites. Yet another tradition regarded Stonehenge as the site of a tomb dedicated to Boudicca, a Celtic queen who had valiantly opposed the Roman conquest of her island. Still others looked to the legendary days of King Arthur and claimed Stonehenge was a memorial to slain knights built by none other than Merlin the magician from stones he had transported from Ireland through levitation.

Serious archaeological study of the site, however, did not begin until the seventeenth century and continues to this day (fig. 47). Before reviewing the results of these investigations and considering their bearing on the history of ancient science, it is vital to understand the complex structure of the site.

Figure 47: Eighteenth-century views of Stonehenge. From Edgar Barclay, Stonehenge and Its Earthworks (London: Nutt, 1895).

Stonehenge consists of three basic components. On the outer rim of the site is a ditch with an earthen bank that forms a circle measuring some 320 feet in diameter. Within that circle there is a ring of holes once dug in the earth and then later refilled. Some of these holes (known as Aubrey holes, after the name of their seventeenth-century excavator) may have once held upright stones or posts, while others were found to contain cremated human remains. Set at intervals along the ring are four "Station" Stones that constitute the corners of an invisible rectangle.

If we continue to move inward toward the center of the site, we come upon an outer circle of sixteen (originally thirty) vertical "sarsen" stones, each standing thirteen and a half feet tall and weighing between twenty-five and fifty tons. On top of them rest seven-ton sarsen lintels. (The name *sarsen* comes from *Saracen*, the name of the enemies of the medieval crusaders and thus a generic English term for foreigner—the "foreignness" of the stones being due to having come from a quarry twenty miles away.) Inside this outer stone circle is an inner one measuring seventy-six and a half feet in diameter. This circle is made of upright blocks of rare bluestone that were hauled over land and sea from a site in Wales some 240 miles away. Within this inner circle stands a horseshoe formation of massive sarsens arranged in groups of three, with two uprights supporting a horizontal lintel. Of these "trilithons," as they are known, only two sets of an original five remain. Finally, within the sarsen horseshoe is a smaller bluestone horseshoe containing a stone originally dubbed (from Druidic legend) the "Altar" Stone. Both horseshoes open toward the northeast and point toward a one-and-a-half-mile-long ceremonial causeway traditionally called the Avenue.

On the Avenue are two special stones: a flat "Slaughter Stone" at the opening of the causeway, which may mark the site of a former gateway, and a giant upright stone called the "Heel Stone." Legend says this one got its name from the time Satan threw a stone at a friar and hit him in the heel. If you doubt the veracity of this legend, there are two other etymological options to choose from: that "Heel" comes from the ancient Greek word for the sun (*helios*); the other, that it is derived from the Saxon word for *hide* (*helan*), because the sun—as we will soon learn—can hide behind the stone.

To review, then, working from the outside in: there is a ditch and an earthen bank; a ring of Aubrey holes with Station Stones; an outer circle of sarsen stones; an inner circle of bluestones; a sarsen horseshoe; a bluestone horseshoe; and finally, to the northeast, an adjoining causeway (plate 13).

As you can tell, Stonehenge is a structurally complex site. Archaeological evidence shows it was not built all at once but was designed and constructed in stages over the course of more than a thousand years. Most archaeologists agree that the ditch and the bank, together with the Aubrey holes and the Heel Stone, probably came first. They are dated to about 3000 BCE, based chiefly on the age of the human remains found in the holes. What components of Stonehenge came next and in what order is a subject of debate. One theory is that the sarsen and bluestone circles appeared next, sometime between 2500 and 2000 BCE, followed by the sarsen and bluestone horseshoes around 2000 BCE. Whatever the exact dates, all the experts agree that Stonehenge was constructed long, long before the days of King Arthur, Boudicca, and the Druids—thereby eliminating them as the monument's likely builders.

Like the work crews of the pharaohs, the builders of Stonehenge lacked iron tools to quarry and sculpt the stones, wheeled vehicles to transport them, or pulleys to hoist them into place; they instead relied on stone-headed mallets, sledges, earthen ramps, levers, and raw muscle power. Unlike the pyramid builders, however, the early Britons, if we are to judge them by their surviving artifacts, were a primitive, albeit determined, folk.

MONUMENTAL MOTIVATION

What motivated these early builders to undertake such a massive and time-consuming project?

One theory is that Stonehenge may have been a monument to the dead. The radiocarbon dating of human remains from the site suggests that burials may have taken place as much as a thousand years before the largest of the stones were erected and continued to take place thereafter. Thus Stonehenge may have served primarily as a cemetery, with massive prehistoric tombstones acting as markers memorializing the dead. Who these dead were we do not know, although we have learned that the artifacts interred with their ashes point to an elite class of rulers or priests.

Another theory—and the two are not mutually exclusive—is that the site functioned as a prehistoric astronomical observatory designed to predict and ritually celebrate major celestial events. This theory was first proposed as early as 1740 by William Stukeley, who noted that the main axis of the monument, that is, the long axis of its horseshoes, points in the direction of the summer and winter solstices (June 21 and December 21).[3] This theory was elaborated in 1901 by Sir Norman Lockyer,[4] and then advanced in 1963 by Gerald S. Hawkins with the aid of an early IBM computer.[5] After programming the computer with the historical positions of the sun, moon, and stars in antiquity, Hawkins entered the alignments of all Stonehenge's structural components. The result: an astounding number of alignments that had celestial significance. The number of matches generated by Hawkins's analysis proved comparable to his having tossed a coin in the air twenty-three times and come up with nineteen heads—far more than chance alone would seem to allow. The main elements in these cosmic alignments were the Heel Stone, over which the summer sun dramatically rose on June 21, and the massive trilithon "gateway" opposite it, through which the setting of the sun on December 21 could be viewed (plate 14). In addition, the fifty-six Aubrey holes may have served as counters that permitted the primitive Britons to calculate the coming of lunar and solar eclipses.

Thus, rather than being just a cemetery, Stonehenge could have also been a sanctuary dedicated to the awesome mysteries of the heavens, a sanctuary that would have given people reassurance by enabling them to come to grips with those mysteries under the tutelage of their learned priests. To appreciate what was in the minds and hearts of Stonehenge's builders, we must divest ourselves of our rationalistic understanding of the universe, imagining instead the fear the early Britons would have felt when the sun or moon vanished from the sky in an eclipse or the sun seemed to be disappearing beneath the winter horizon as the days grew colder and shorter. Or, imagine the relief they would have felt when the days once again blessedly lengthened and new crops could grow. At those times, the site of Stonehenge may have served as a theater where thousands of spectators would have watched in awe as a celestial drama unfolded.

Stonehenge is not unique. Some nine hundred smaller prehistoric monuments lie scattered across the British Isles, many with their circular shapes still largely visible. At nearby Avebury stood a circle five times Stonehenge's diameter, though farmers long ago broke down and carted away most of its megaliths to clear the way for their pasturing herds. Across western Europe, especially on the coast of Brittany at Carnac, other such monuments still stand in mute testimony to a cosmic consciousness continental in scope.

Because *prehistoric* means "before writing," these ancient sky watchers, unlike the Greeks, have left behind no astronomical commentaries—indeed, no literature at all that could illuminate their beliefs. The stones themselves are their only witness. Indeed, apart from a simple understanding of mechanics, astronomy was their only science. And even this claim to fame has been challenged by some who say a people so materially backward could never have been so intellectually advanced.

Almost a century ago, H. G. Wells, the author of *The Time Machine*, wrote an essay titled "The Grisly Folk." In it he envisioned a future time when human beings like ourselves might be able to intuit the thoughts and feelings of our prehistoric kin. He wrote: "A day may come when these recovered memories may grow as vivid as if we in our own persons had been there and shared the thrill and the fear of those primordial days; a day may come when the great beasts of the past will leap to life again in our imaginations, when we shall walk again in vanished scenes, stretch painted limbs we thought were dust, and feel again the sunshine of a million years ago."

Chapter 18

SCIENCE IN ANCIENT CHINA

Among the ancient civilizations of the Mediterranean, the Greeks were not the first to seek a deeper understanding of themselves and the world. As we have seen, the thinkers of Egypt and Mesopotamia had been there before, and the Greeks carried on their quest, but with a new and unwaveringly pure and passionate commitment to the promise of reason. However, long before the civilization of the Greeks blossomed, another civilization had already flowered far beyond the Mediterranean, the civilization of ancient China—the third to arise among the world's great civilizations. From the genius of ancient China would come a host of scientific discoveries.

A COSMIC TALE

Let us begin with a Chinese legend and follow its path through the long and winding history of that great land.

Borne aloft on the wings of a dragon, Ch'ang O ascended skyward until she came to a great and luminous sphere. It was the moon, symbol of the *yin*, the feminine principle of creation. Its landscape was desolate and cold. Only the cinnamon tree grew there, for nothing else could flourish.

Down on Earth, her husband, Shen I, was angry with her. Rewarded long before with a pill that could render him immortal, Shen I had hidden it in the rafters of their home. But one night when he was away, his wife had seen a beautiful white light coming through the ceiling and, following it, had found and then swallowed the magic pill. When he returned home and found his treasure gone, Shen I flew into a rage. Fearing his wrath, Ch'ang O fled and was transported into the sky.

Later, Shen I learned that he, like his wife, could become immortal. He could do so by riding on the back of a golden bird to the sun, the symbol of the *yang*, the masculine principle of the universe. After his successful trip, he traveled on to the moon on a ray of sunlight, for he still loved his wife and had grown lonely for her.

Once on the moon, he took the wood of the cinnamon tree and built Ch'ang O a palace, promising to visit her on the fifteenth day of every month. That is why the moon is full on that night, for the moon then welcomes the full radiance of the sun as in harmony yang and yin become one.

THE LADY IN RED

Four hundred miles south of Beijing, twin mounds stood for over two millennia, bathed in the full moon's light. Beneath one lay the bodies of a Chinese nobleman and his son; beneath the other, the body of the nobleman's wife. They had lived in the second century BCE, a century after the reign of China's first emperor. The discovery of their tombs in 1971 was to disclose the details of their lives and their quest for eternal life.

The nobleman's wife was buried in the eastern mound, the place of greater honor. When she died at the age of fifty, the attendants washed her skin and coated it with a secret ointment to preserve her flesh. Servants combed her hair and fastened her wig in place. They then dressed her in twenty layers of silk fastened with silken cords, fitting silken slippers upon her feet.

A sepulcher was dug, floored in white clay, and spread with bamboo matting. Three painted coffins, one inside the other, were to hold her body; three wooden boxes, one within the other, would enclose the coffins. Over the boxes five tons of absorbent charcoal were laid down and, over that, white clay two to four feet thick. Finally, red clay was heaped up as the final barrier to moisture and decay.

Food for banqueting was also put in the tomb, the complete stock itemized on a bamboo menu. There was an exotic array of meat, fowl, and fish (pork, mutton, venison, beef, hare, and dog; chicken, duck, wild goose, pheasant, crane, turtledove, and sparrow; carp, bream, and mandarin-fish). Fruits like plums, pears, dates, and red bayberries were provided, as were condiments such as honey. A 180-piece service of glistening dinnerware coated in orange and black lacquer was packed with accompanying sets of chopsticks.

To attend her ladyship were servants in the form of dolls that would come to life in the hereafter. The dolls were carved out of wood and painted, the larger ones—almost three feet tall—were dressed in silk. Among the 162 dolls were ladies-in-waiting, dancing girls, and musicians, some holding woodwinds, others kneeling before miniature zithers.

The lady's love of music (perhaps her longing to play) was satisfied by the inclusion of full-sized instruments, among them a mouth organ made of twenty-two bamboo pipes and a zither almost four feet long with tuning pegs, moveable bridges, and twenty-five silk strings.

During her lifetime, such music would have provided succor from pain. Years before—as an archaeological autopsy would reveal (plate 15)—she had contracted tuberculosis, and her lungs still bore the scars. Large stones blocked her gallbladder ducts. Her intestines harbored parasites. Her back ached from lumbago. And arteriosclerosis, so advanced that her left coronary artery was nearly blocked, was to bring about her death.

Her physician, anticipating that she might feel discomfort even in the afterlife, laid packets of herbal medicine in her tomb—peppercorns, magnolia bark, and magic cinnamon—remedies used in traditional Chinese medicine to treat heart disease to this very day.

The scientific ingenuity that had been used to protect her body had not been in vain. When extricated from her multiple coffins, her body showed no sign of decomposition. Unlike the hardened and brittle mummies of ancient Egypt, her yellowish-brown skin was still elastic, her flesh supple after twenty-one centuries. Even the joints of her limbs were still pliable.

This remarkable preservation can no doubt be attributed in great measure to the sepulcher itself; hermetically sealed, it had barred oxygen, moisture, and bacteria. But the body's condition can also be explained by chemical treatment: traces of a curious reddish fluid (a mercury compound mixed with organic acids) was found on the body as well as on some of the objects in the tomb.

Because of the body's preservation, a modern examination was able to disclose the details of the lady's medical history. In addition to determining the probable cause of her death, the autopsy even revealed what she had been eating just prior to her heart attack: in her intestines, stomach, and esophagus were found 138 and a half melon seeds.

From inscribed seals found in her tomb and in the companion tomb of her husband and son, we know that this ancient Chinese noblewoman was the wife of Li Ts'ang, the Marquis of Tai. Another seal may even carry her own name: Lady Hsin Chui.

Of all the discoveries in her tomb surely the grandest is a seven-foot-long piece of silk embroidery. Known as a *fei i*, or "flying garment," it was draped over her innermost coffin. Shaped like the letter *T*, it resembles a long kite.

Embroidered on it we see Lady Hsin Chui, bent with age and infirmity, followed by her servants. Below are creatures of the mythic underworld; above, the sun's orange orb. High in the sky, rising on the wings of a dragon, is Ch'ang O, visitor to the moon, ascending into the lunar light. She seems so delicate and young, so very young, even as Hsin Chui had once been. Perhaps Lady Hsin Chui imagined that, when dead, she, too, might fly, her soul wafted to heaven by the kite, rippling through the wind, the cinnamon moon and silken immortality within her reach.

Two years after the tomb of Lady Hsin Chui was opened, archaeologists excavated the companion mound containing the graves of her husband and son. The tomb of the marquis had been robbed more than once. Apart from scattered seals bearing his name, little of value was found. The grave of their son, however, had not been plundered. Here were lacquerware and lengths of silk, foods to be feasted on in the afterlife, and musical instruments to be played.

Most fascinating among the finds were manuscripts made of silk or strips of bamboo, placed in the tomb because they had been treasured possessions during his life and could provide nourishment for his mind during the endless hours of eternity. Among them were the oldest maps of China ever found, military maps showing topographical features, roads, and cities, and indicating the size of populations and the location of military garrisons. Discovered also were the oldest-known works of Chinese astronomy, treatises so advanced that they gave the orbital periods of five planets: Mercury, Venus, Mars, Jupiter, and Saturn. There were classics of Chinese philosophy too: the sixth-century BCE meditations of Lao Tzu and the centuries-older *I Ching*, or *Book of Changes*, explaining the doctrine of yin and yang. Warfare, science, and philosophy had all been concerns of the prince while he was alive. To keep him healthy in the afterlife, medical texts were buried with him. Healthful bending and stretching exercises were illustrated in cartoon-like fashion on silk, and ways to prolong life were described, including the avoidance of cereals, denigrated as the food of peasants.

For over two millennia, the moon had bathed the burial mounds in its light. By their wealth and station, like the pharaohs of Egypt, three aristocrats had contrived to gain immortality. Beneath the same undulating landscape, the simple graves of millions whose lives were measured by bowls of precious rice still lie in undisclosed anonymity.

THE ROOTS OF WISDOM

The contents of the tombs we have just explored bear witness to traditions of science and technology deeply rooted in China's cultural past.

The lacquer coating on our aristocrats' dinnerware is the world's oldest example of an industrial plastic, made from a sap extracted from the lacquer tree (*Rhus verniciflua*), a tree indigenous to China that was used for this purpose as early as the thirteenth century BCE. Patiently applied in multiple (sometimes even a hundred or more) thin coats, this natural plastic made Chinese dinnerware not only colorful but impervious to heat. A chance event—crab meat falling into a paint pot!—led artisans to discover a chemical substance in the crab's tissue that could inhibit the lacquer from hardening prematurely. As we have already seen, chemical expertise was also applied with similar softening intent to keep the flesh of the corpse supple for all eternity.

The musical instruments interred in these tombs testify not only to a love of music but to a knowledge of harmonics. As far back as the sixth century BCE, the Chinese were casting bronze bells in different sizes and shapes to give them distinct tones that would enable them to be played in unison, even going so far as to strategically engineer bells with two different "strike points" to produce two different tones. The tomb of one fifth-century BCE nobleman housed a total of sixty-four bronze bells to be played in concert, the largest of which was five feet tall and weighed 447 pounds. Ancient Chinese musicologists realized that lower notes were produced by slower vibrations, and higher ones by quicker ones; they also deduced, from the behavior of a zither's strings, the nature and effect of sympathetic vibrations.

The herbal remedies and the health regimen prescribed for Lady Hsin Chui and her son in this life and the next are testimony to the science of medicine in ancient China. By the second century BCE or earlier, Chinese physicians recognized that the blood, pumped by the heart, circulates throughout the body through arteries, veins, and capillaries. At the same time, they held that a form of vital energy called *ch'i*, pumped by the lungs, circulates through corporeal channels that are invisible. The blood they identified with yin, and the vital energy with yang, and they viewed health as the product of their cooperative and harmonious interaction. An imbalance between yin and yang, they believed, could be restored through the use of acupuncture, a practice referred to as early as the sixth century BCE and whose origins may reach back another thousand years. Indeed, in a tomb dating to the late second century BCE, archaeologists uncovered a set of nine gold and silver acupuncture needles whose traditional number and shapes point to their therapeutic use.

Also, by the second century BCE, Chinese medical investigators had identified sex and pituitary hormones, which they isolated from human urine and used for the treatment of medical conditions. Around the same time, they also noted how the body is more susceptible to certain diseases at certain times of the day or night and at certain seasons of the year—thus anticipating the discovery of the human body's invisible circadian rhythms. By about 200 CE, Chinese physicians had concluded that many medical conditions are caused by dietary deficiencies and, though lacking a knowledge of vitamins, prescribed precisely those foods that were rich in the very nutrients that were needed to restore a patient's health. By the third century CE, they were using an extract of mandrake as an effective anesthetic for surgery, administering it in an effervescent form when stirred into wine.

Silken kites, like the one buried with Lady Hsin Chui, had a long history. Inspired by a longing to fly upward toward heaven like the birds, the Chinese constructed kites, often in the shape of birds, as early as the fourth century BCE, and learned about aboveground air currents from their hands-on experience. Among the most famous kite makers was the third-century BCE moral philosopher Mo Tzu, who is said to have devoted three years to building a special model. By the fourth century CE, a famous alchemist named Ko Hung was even speculating about making a kite large enough to carry a man by employing the principle of aerodynamic lift. Alas, in the following century, a sadistic emperor named Kao Yang used his enemies as "test pilots" for kites by having them jump off a hundred-foot-tall tower. As the story goes, when one prisoner actually succeeded in gliding two miles, he was imprisoned without food until he died. Earlier aeronautical experiments in the second century BCE included making miniature hot-air balloons (by igniting bits of tinder inside hollowed-out eggshells) and devising makeshift parachutes (by tying together a sufficient number of straw coolie hats).

Beyond the sky, of course, were the heavens themselves. The tale of Ch'ang O and her trip to the cinnamon moon is but a mythic metaphor for the efforts of Chinese astronomers who, unaided by telescopes, investigated the moon and the other mysterious celestial bodies that lay beyond. Unlike the Greeks, however, their work was motivated not by a desire to understand the cosmos for the intellectual satisfaction it could provide, but to predict the future in the belief that the destiny of humankind was written in the stars.

Like the talents of the craftsmen whose handiwork filled the tombs of nobles, the activities of astronomers were governed by China's emperors, who believed accurate horoscopes could inform their policy making and military planning. Indeed, they viewed the making of horoscopes as a matter of national security. The by-product of such pragmatic astrological research, however, was the acquisition of a vast store of astronomical knowledge. Because of the continuity of China's cultural practices, which have persisted for over three thousand years, "the Chinese possess the longest unbroken run of astronomical records in the world, observations of considerable importance to modern astronomers, whose research requires data about long-term celestial events."[1] In particular, Chinese astronomers were fascinated by eclipses and the dramatic appearance of comets. Their records, in fact, document a solar eclipse in 1302 BCE, the arrival of Halley's comet in 240 BCE, and the sighting of supernovas.

The stories of Chinese astronomy and Chinese medicine intersect when it comes to a group of artifacts known as "oracle bones." Made from the shoulder blades of oxen (or, in some cases, from the shell underbellies of tortoises), oracle bones are inscribed with prophecies, not unlike modern-day fortune cookies—only these "fortune cookies" date as far back as 1300 BCE and bear some of the very earliest examples of Chinese writing. For many centuries, because of their antiquity, they were ground up and used for potions in traditional medicine, where they were known as "dragon bones," that is, until in 1899. A Chinese scholar that year detected in his pharmacy prescription some primitive types of Chinese pictograms. They were visible on some larger bone fragments that had escaped the pharmacist's mortar and pestle. This discovery led archaeologists to their source, a buried palace archive at a place called Anyang, some three hundred miles southwest of Beijing, where they uncovered a cache of over 150,000 such bones. Most were inscribed with questions that had been put to departed ancestors, together with the "yes or no" answers that supposedly came from the spirit world. The fortune tellers obtained the answers by drilling minute holes in the bones, heating them with hot metal pins, and tracing and interpreting the cracks induced by the heating process. The inscriptions were dated with celestial coordinates and recorded unusual celestial events that had been observed at the time of the reading.

Back on Earth, Chinese geographers mapped their world, in large part—as the tomb of Hsin Chui's son suggests—for military purposes. By the third century BCE, they were drawing the world's first relief maps or carving them in jigsaw-like fashion out of wood, sculpting their three-dimensional features out of sawdust, flour, and water. A first-century CE general even wrote a treatise titled *The Art of Constructing Mountains out of Rice*!

WONDERS OF THE ORIENT

The Chinese used their ingenuity to invent other devices with distinct military applications, most notably the rope suspension bridge (first century CE) for crossing rivers and gorges, and the mechanical crossbow (fourth century BCE), that extended the range and impact of arrows that would otherwise have been fired by hand. Though the first compass with a magnetized needle would not be invented in China until the eleventh century CE (nevertheless a full century before its first European counterpart), precursors made of magnetic lodestone (plate 16) existed as far back as the fourth century BCE and were employed not for military or navigational purposes but to magically align the construction of houses and cities with Earth's hidden spiritual forces.

Perhaps the most extraordinary invention of all was the so-called south-pointing chariot of the third century BCE (plate 17), an invention worthy of those master Hellenistic engineers, Heron and Ctesibius.[2] As a permanent passenger, the two-wheeled chariot carried a statue that could rotate 360 degrees on its base. Once set, it would always point south (and away from the barbarous northern frontier), no matter which way the winding roads of China took it. What made the statue work was a differential gear that detected the difference in the two wheels' rotation when they made a turn and then pointed the extended arm of the statue in exactly the opposite direction. It would not be until 1879 that a differential gear made its debut in a European vehicle.

Other marvelous Chinese devices included metal bowls that could spout water when their handles were rubbed (fifth century BCE); water vases designed with off-center "centers" of gravity that caused them to tip over when they were empty (third century BCE); magic lanterns, or zoetropes, that projected images of moving animals onto their screens (second century CE); a seismograph that dropped a bronze ball into a bronze toad's mouth whenever a tremor occurred, however faint and far away it might be (second century CE); and a magic "see-through" mirror made of solid metal that, when held in the sunshine, allowed one to look at the smooth and shiny surface of its front and see a hidden image engraved on its back (fifth century CE). No doubt all of these marvels amused and delighted the nobles and emperors for whom they were made.

BEYOND THE GREAT WALL

The Great Wall of China (fig. 48) was a colossal monument to a monomaniacal dictator's ego.

Warlord at the age of thirteen, Ying Cheng battled the other warlords of his land until he had crushed them all, leading a million soldiers into battle until he had devoured his enemies like a silkworm feasting on mulberry leaves. From the melted metal of their weapons, he fashioned gigantic effigies of himself, took a newly created title, "Divine Emperor," and prophesied that his dynasty would rule for ten thousand years. It would last but a generation.

Ying Cheng's most enduring monument was a wall built by the forced labor of prisoners of war and of Confucian convicts whose subversive thoughts the emperor feared. If we believe the tale, their bones were ground up and added to the wall's mortar, making it the world's longest cemetery. Constructed in the third century BCE by linking together older feudal walls, it stretched for fifteen hundred miles from the Gobi Desert to the Pacific—indeed, twice that length if we straighten out its dragon-like twists and turns. Planned by China's first emperor as a line of defense against northern invaders, it also symbolically served as a line of demarcation to define his continental domain.

Figure 48: View of the Great Wall of China. From Raphael Pumpelly, Across America and Asia: Notes on a Five Years' Journey Around the World and of Residence in Arizona, Japan, and China *(New York: Leypoldt & Holt, 1871).*

The design of the Great Wall testifies to methodical and scientific planning. Surmounted by a paved pathway wide enough for eight men to march abreast, it included a series of twenty-five thousand watchtowers, each twenty feet tall, located two bowshots apart to provide its defenders with complete ballistic coverage. The towers also doubled as communication stations from which coded messages in fire or smoke could be transmitted across the wall's entire length in twenty-four hours.

Ying Cheng expended similar energy in planning and executing the construction of the tomb meant to guarantee his immortality. Shaped like an earthen pyramid to replicate the zenith of heaven, it rose to a total height of fifteen stories and towered over a temple-filled spiritual "city." To protect the emperor in the afterlife, his tomb was surrounded by a buried army of larger-than-life, baked-clay soldiers, sculpted and painted in exquisite detail and numbering almost ten thousand strong. Arrayed with their commanders in full military formation, they were accompanied by wooden chariots drawn by terra-cotta horses and carried wooden crossbows whose metal triggers and arrowheads were unearthed in the 1970s, along with the ghost army itself.

Tradition tells us (for painstaking archaeologists have not yet entered the tomb) that the emperor's burial chamber was filled with marvels. According to a history written just a century after the emperor's death, the tomb contained models of palaces, precious stones, and other rarities. Perhaps the most extraordinary treasure was an ingenious mechanical model of the entire empire, with miniatures of the Yellow and Yangtze rivers made of mercury that flowed in perpetual motion into a mercury ocean set beneath a bejeweled sky. The body of the emperor himself rested in a copper coffin in a chamber sealed by a door encrusted with jade—a substance the ancient Chinese regarded as more precious than gold. The door's iron frame was magnetized to trap and hold any armed intruders, assuming they had not already been struck and killed by automatic crossbows triggered by their footfall.

Someday, no doubt, archaeologists will excavate the red earthen mound beneath which the corpse of cruel Ying Cheng still lies.

"Perhaps the strings of the automatic crossbows hidden inside have long since rotted away. Perhaps the copper coffin has corroded, green beyond recognition. Or perhaps somewhere in the still darkness, jeweled stars are waiting to sparkle once again, and rivers of quicksilver to shimmer once more as they course to an endless sea."[3]

Figure 49: Marble head of a companion of Odysseus. © The Trustees of the British Museum; Art Resource, NY.

EPILOGUE

THE "UNTRAVELL'D WORLD"

"Yet all experience is an arch wherethro
Gleams that untravell'd world, whose margin fades
For ever and for ever when I move."

—Tennyson, *Ulysses*

The ancient Greeks were a race of explorers—exploring the world in all its infinite wonder and exploring the wondrous inner universe of man himself. Like Homer's Odysseus, the inspiration for Tennyson's poem, they were driven by a restless curiosity to discover what lay beyond the horizon (fig. 49, at left).

Our own journey through the course of these pages has revealed the marvelous breadth and depth of those ancient explorations, ranging far beyond those undertaken by other early civilizations, however marvelous or sophisticated their individual accomplishments may have been. Inspired by a sense of wonder and guided by the penetrating light of reason, Greek thinkers investigated the natural world, forever asking fundamental questions about its substance and design. In their quest for answers, they traveled the earth to its farthest limits and scanned the outer reaches of the heavens. With equal intellectual dedication and passion they probed themselves, examining the hidden workings of both body and mind.

Though ancient Greek science is the progenitor of modern science, it differed from it in three major respects.[1] First, it relied more on formulating theories in the abstract than on developing them through hands-on experimentation and observation. In effect, with the notable exception of biology, Greek science took place more in the "study" than in the "lab" or "field," and tended to be more deductive (moving from preconceived premises to particulars) than inductive (moving from particulars to generalized conclusions). That was both its grandest strength and its greatest weakness.

Second, Greek science was mostly performed by ordinary citizens working in their leisure time on inexpensive projects, whereas today it is carried out by highly degreed employees of universities, governments, or corpo-

239

rations who rely on salaries and grants to finance costly research. This difference meant that the ancients had more economic independence and intellectual autonomy than do their modern counterparts.

And last, except for the inventions of the Hellenistic period, most ancient Greek science was pure rather than applied. As such, the research conducted by Greek scientists was much less focused on potential military or commercial applications than it is today, and so posed fewer moral dilemmas. Though the practice of slavery throughout the classical world was recognized by the Stoics as inhumane, it persisted. And, by providing a ready supply of manpower, it inhibited ancient scientists from developing those new sources of energy and types of technology that since the Industrial Revolution have, for better *and* worse, transformed our lives.

All three of these differences persuasively demonstrate the predominant role of intellectualism in Greek culture and the Greeks' exceptional fascination with the life of the mind, qualities that largely set them apart from their ancient predecessors and contemporaries, and qualities that are predominantly absent from the culture of today.

Though the specific discoveries of the Greeks have been described earlier in detail, far more important than any list of discoveries is the underlying attitude that explains why they were made in the first place. It is this attitude, the "scientific attitude," that remains the most visible contribution of ancient Greek civilization to the modern world, a contribution that stands as its most influential living legacy.

Committed to the search for truth, Greek scientists sought logical rather than supernatural answers to life's questions. They used the language of mathematics instead of the language of prayer. In addition, they believed that human beings must be free to seek out those answers rather than let their imaginations be shackled by the authoritarianism of religion or government. Convinced, like Socrates, that the unexamined life isn't worth living, they took the power of the human mind seriously, as seriously as they did the value of freedom. Consequently, almost nothing for them was off-limits.

There is a lesson to be learned from the Greeks.

The future of science on our planet will depend upon our keeping faith with their uncompromising attitude. To the extent that objective truth is muffled by political policy, stifled by religious dogma, gagged by profiteering, or suffocated by public indifference, to that extent we will betray not only the Greeks but our very own capacity to be a better nation. We must therefore continue to explore "that untravell'd world," for nature's mysteries guarantee there will always be a road left that remains untraveled. Like Tennyson's Ulysses, then, we must "follow knowledge like a sinking star / Beyond the utmost bound of human thought."

The Greeks, however, in particular their playwrights, had few illusions about their own omnipotence or omniscience, and when they did, it cost them dearly. Celebrating humanity's achievements, Sophocles once wrote:

> *Many are the world's wonders, but none more wondrous than man.*
> *Under the south wind's gale, he traverses the gray sea,*
> *knifing through its surging swells.*
> *Earth, eldest of the gods, imperishable and everlasting,*
> *he erodes year after year with winding furrows*
> *cut by his equine team.*
> *The winged flocks of birds,*
> *the wild herds of beasts,*
> *and the salt-sea schools of fish*
> *he entraps in the woven mesh of his devious net.*
> *With his devices he overpowers the creatures of the wild,*
> *reining in the shaggy-maned stallion*
> *and yoking the stubborn mountain bull.*
> *Speech he developed and wind-swift thought*
> *and the talent to dwell together, and learned*
> *how to evade the chilling frost and pelting rain.*
> *Ingenious, there is nothing that comes he cannot master.*
> *Only from Death can he not contrive an exit*
> *though from contagion he has made his escape.*
> *With a brilliance and subtlety beyond imagining*
> *he gravitates at times toward evil, at other times toward good.*[2]

As Sophocles's *Antigone* makes clear, for all his wondrous talents, man can become his own worst enemy when he acts with hubris and fails to anticipate the potentially tragic consequences of his choices.

Today we possess far more knowledge and power than the ancient Greeks ever did. But it is precisely because of this that we must hearken all the more closely to their warnings about the human propensity to err catastrophically. This, indeed, may constitute our greatest scientific and technological challenge in the years ahead: to courageously search for truth as did the Greeks, but not transgress the moral boundaries that define our very humanity.

THE HELLENIC
"HALL OF FAME"

Note: Within any given century or group of centuries, individuals are listed alphabetically by name.

Fifteenth(?) Century BCE

Daedalus. The mythical designer of the Cretan Labyrinth and the builder of robotic statues, Daedalus, together with his young son Icarus, became the first man to fly by making wings out of branches, feathers, and wax and taking off from the roof of King Minos's palace. Though Daedalus and his son may have been fictional, the archaeological discoveries of Sir Arthur Evans and his successors have established the historicity of the world to which they are said to have belonged.

Seventh to Sixth Centuries BCE

Thales of Miletus. Ranked first among their "Seven Sages" by the Greeks, Thales was the first thinker ever to speculate on the workings of the universe and to explain its operation in natural rather than supernatural terms. According to tradition, he was the first Greek to accurately predict a solar eclipse and to determine the precise time of the summer and winter solstices. He was the first Hellenic thinker to employ diagrams to represent abstract geometric principles and, during a stay in Egypt, he used geometry to calculate the height of a pyramid by comparing the length of its shadow to the shadow of an adjacent vertical rod. Thales was also the first Greek scientist to investigate magnetism.

Sixth Century BCE

Anaximenes. Recognizing that breath was necessary for life, the philosopher Anaximenes proposed that air was the universe's most essential element. His observation of condensation and combustion led him to conclude that air was also the most versatile of elements since, under the right conditions, it could become water or fire. It is upon a cushion of air, he asserted, that Earth floats.

Pythagoras. The founder of a mystical commune in Italy and—according to tradition—the originator of the word *philosopher*, Pythagoras taught that the universe contained a hidden harmonious order and that the cosmos, like Earth itself, possessed a perfect, spherical shape. While performing experiments with sound, he discovered that the pitch of a musical instrument's string is directly proportionate to its length, an intimation to him of a stunningly larger truth: that all physical phenomena express fundamental mathematical relationships that can be discerned by the human mind. His search for such numerical correlations in elementary shapes led him to the discovery of the right-triangle theorem that still bears his name. Before establishing his commune in Italy, Pythagoras may have visited and studied in Egypt and Babylonia.

Sixth to Fifth Centuries BCE

Anaximander. He was the first to propose that the cosmos was infinite and that the starry heavens above were part of an all-encompassing sphere that embraced Earth. To account for the fact that the locations of stars seemed to change as he traveled, he proposed that the surface of Earth was curved. Anaximander also became the first scientist on record to draw a map of the known world.

Hecataeus of Miletus. The earliest-known scientific historian, Hecataeus wrote the first systematic and descriptive account of the world's geography and its inhabitants. Though Hecataeus's own works survive only in fragmentary form, the literary precedents he set inspired the later historical and geographical researches of Herodotus, which endure to this day.

Heraclitus. The philosopher Heraclitus believed change was central to reality—that change was, in fact, the only constant we can depend upon. Such change was the product of a dynamic tension that pervaded both the universe and the human soul, or *psyche*.

Parmenides. In contrast to the view of **Heraclitus** (above), the philosopher Parmenides regarded the perception of change as an illusion perpetrated upon us by a sensory system that cannot be trusted.

Xenophanes. From the seashells he found on the slopes of mountains, Xenophanes concluded that Earth's surface must have undergone transformation over the course of time. He ridiculed the Hellenic belief in anthropomorphic gods, holding that such gods were shallow projections of man's egotism. He argued instead the vast universe must be governed by a cosmic god whose thinking is far superior to our own.

Fifth Century BCE

Agatharchus of Samos. A painter of realistic scenic designs for the Athenian stage, Agatharchus's book describing his technique inspired Greek scientists to investigate the mysteries of optical perspective.

Alcmaeon of Croton. A physician who specialized in the treatment of the eye, Alcmaeon performed the first-known dissections of the human body and the earliest-known surgical operations on the eye. He identified the optic nerve and the Eustachian tubes that connect the ears to the mouth. Believing that the eyes and other sense organs were connected by passageways to the brain, he correctly identified the brain as the body's operational center.

Anaxagoras. Because of his belief that the universe is composed of four material elements (earth, air, water, and fire) united by abstract intelligence rather than being a creation of the gods, Anaxagoras was charged with impiety by an Athenian court and only barely escaped the death penalty. In the same rationalistic way, he had explained solar and lunar eclipses and the phases of the moon as due to celestial mechanics rather than divine agency.

Empedocles. Empedocles originated the theory, later expanded by **Anaxagoras** (see above), that the universe is composed of four primal "roots" (earth, air, water, and fire). To explain the creation and dissolution of matter, he proposed that these basic elements are alternately attracted (by Love) and repelled (by Strife). Believing that the heart must be the center of the circulatory system, he viewed it as the body's main organ. According to Empedocles, visual images were transmitted to the eye directly from objects themselves rather than, as others claimed, being formed from rays that emanated from the eye.

Herodotus. Herodotus, the "father of history," traveled the known world and, with a keen eye and discerning intelligence, described the customs of its peoples, as well as the flora, fauna, and physical features of the varied lands they inhabited. Despite the fact that he loved to pass along tall tales he had heard on his travels, he also exercised critical intelligence while doing so, often expressing his own doubts as to their veracity or reserving judgment. He left it to his discerning audience to decide what was actually true. His information-gathering method is best illustrated by the etymology of the word *history*, which in Greek originally meant "research."

Hippocrates of Chios. The most esteemed mathematician of the fifth century BCE, he wrote the world's first textbook of geometry, about two centuries before the days of **Euclid** (see below). In the field of astronomy, he is best remembered for his efforts to explain comets.

Hippocrates of Cos. The "father of medicine" and composer of the Hippocratic Oath. His oath is still sworn to by physicians today. Hippocrates founded an influential medical school on Cos, the Greek island of his birth. He believed and taught that diseases originated from natural and not from supernatural causes and needed to be cured not by spiritual rituals but by physical remedies. Agreeing with Empedocles (see above) that sickness resulted from an imbalance among the body's four main fluids, or "humors," he provided a regimen for good health that included attention to hygiene (for both doctor and patient), a moderate diet, and adequate rest to enable the body to regain its internal equilibrium.

Leucippus. Originator of the atomic theory, Leucippus argued that at its most fundamental level matter was composed of minute and indivisible (*atomoi*) particles. He was the first to assert that every event in the natural world has a natural explanation. Leucippus's most famous student was **Democritus** (see below).

Meton. Curious about the mathematical correlation between the solar year of approximately 365 days and the lunar year of about 336 (28 days × 13 lunations), this Athenian astronomer discovered that the two came out virtually even after the passage of nineteen solar years (the so-called Metonic cycle). He then devised a universal calendar combining solar with lunar elements. To this day, the Metonic cycle is used to calculate the date of Easter.

Oenopides. This astronomer calculated the tilt of Earth's axis in its orbital plane (the "obliquity of the ecliptic") and measured the length of the solar year with greater precision than had ever been done.

Philolaus. The most popular exponent of Pythagorean philosophy, Philolaus may have been the first to suggest that Earth moves through the void of space.

Zeno. Like his elder contemporary **Parmenides** (see above), the philosopher Zeno regarded change as a misleading mirage foisted upon us by our faulty senses. To drive his point home and entertain his audience at the same time, Zeno used paradox to show, for example, that motion is a practical impossibility. Fleet-footed Achilles can never catch a slow tortoise, Zeno wittily observed, because every time Achilles reaches the spot where the tortoise was, the tortoise has already crawled farther!

Fifth to Fourth Centuries BCE

Democritus. Elaborating on the theory first proposed by his teacher **Leucippus** (see above), Democritus taught that all things were composed of tiny, indestructible atoms clustered in distinct patterns. Rather than responding to the capricious will of the gods, the operation of these particles obeyed immutable and impersonal laws of nature. Light, he asserted, consisted of atoms in transit. The physics of Democritus was later elaborated in philosophy by **Epicurus** (see below) and in poetry by the Roman author Lucretius.

Plato. A student of Socrates and himself one of Greece's greatest philosophers, Plato proposed that the human personality consists of three basic parts (appetite, will, and reason), each of which must coexist in harmony, with appetite and will playing their proper roles under the guidance of reason. Differing from **Empedocles** (see above), Plato held that vision was a form of energy that flowed from the eye to its object and then returned to the eye. He taught that the cosmos embodied mathematical perfection and that the study of mathematics was the pathway to appreciating its ideal beauty. Plato's greatest student was **Aristotle** (see below).

Fourth Century BCE

Archytas. A Pythagorean philosopher and mathematician, Archytas is credited with founding the science of mechanics. Investigating the structure of music, he proposed that the pitch of a sound was a function of its vibrational speed, or frequency. In addition, he explained the nature of musical scales in mathematical terms.

Aristotle. A student of Plato and a master logician, Aristotle possessed acute powers of observation, a talent for organization, and a wide-ranging intellect attracted to the various facets of the physical world. In the field of biology, he became the "father of zoology," collecting hundreds upon hundreds of specimens and classifying animals by type in an ascending ladder of increasing complexity. In the field of mechanics, Aristotle discussed the effects of force and inertia, and was the first to document robotic mechanisms he had viewed in operation. Starting with Empedocles's four fundamental elements, Aristotle added a fifth and celestial element he called "ether." Of all the Greek philosophers, his impact on the thought of the later Middle Ages was the most profound.

Dicaearchus. Renowned for his intellectual versatility and knowledge, Dicaearchus constructed a map of the known world stretching from the Strait of Gibraltar to the Himalayas. It was marked with a parallel of latitude that would allow any observer along its length to view the sun from exactly the same angle as any other such observer.

Diocles. After **Hippocrates** (see above), Diocles was the second-greatest medical expert of ancient Greece. He is said to have authored the first handbook of anatomy, a word he may have invented. In addition, Diocles introduced innovations into surgery and postoperative care. As a medical educator, he stressed the importance of a physician's practical experience and the need to treat patients as individuals.

Eudoxus. The second-greatest astronomer of Greece (after **Hipparchus**), Eudoxus divided the heavens by drawing lines of latitude and longitude, in effect creating a celestial globe. Discovering that the apparent altitude of stars changed with the latitude of the observer, Eudoxus provided convincing evidence that the surface of Earth is curved.

Heraclides of Pontus. He proposed that, rather than the vault of heaven turning overhead at night, it was the rotation of Earth itself that could be responsible for the apparent motion of the stars. The rotation of Earth on its axis could likewise account, he said, for the alternation of night and day.

Praxagoras. This Greek physician was the first to distinguish between veins and arteries. He concluded, however, that only veins carried blood and that arteries were filled with air (as it appeared from autopsies). He also studied the connections between the spinal cord and the human brain.

Pytheas. On a voyage of exploration into the Atlantic, this Greek navigator sailed as far as the northern coast of Scotland, and even perhaps to Norway or Iceland, taking astronomical observations along the way to determine his latitude. He recognized that the North Star is not stationary but that it moves in a small circle during the course of the night. Pytheas also theorized that the ocean tides he witnessed were caused by the pull of the moon and were affected by its phases.

Fourth to Third Centuries BCE

Aratus. This poet wrote an ancient "best seller," *The Phenomena*, inspired by an astronomical treatise written by **Eudoxus** (see above). In addition to disquisitions on the stars, the poem discussed signs in nature that warned of bad weather.

Aristarchus of Samos. Contrary to the humanistic notions of his countrymen, Aristarchus set forth the unpopular theory that the sun, not Earth, is the center of the universe and that all planets, including Earth, revolve around it. Though he lacked the instruments for taking precise sightings, Aristarchus used geometry to calculate the relative distances of the sun and moon from Earth. Furthermore, he used the shadow that Earth cast on the moon to estimate the moon's size and, by extension, the size of the sun. In the end, it was **Ptolemy**'s flawed geocentric model of the universe (see below) rather than Aristarchus's correct heliocentric model that dominated Western scientific thinking down to the days of the Renaissance.

Ctesibius. This mechanical genius invented a self-regulating clock and a musical organ, both powered by water, and an air-powered catapult. Though his own writings no longer survive, his inventive legacy was kept alive in descriptions by **Philon** (see below), **Heron** (see below) and the Roman architect and engineer Vitruvius.

Herophilus. This anatomist dissected the human eye; distinguished between the cornea, iris, and retina; and identified the optic nerve, tracing it to the brain, which he regarded as the operational center of the nervous system. In addition, he followed the track of the human circulatory system to and from the heart, investigated the diagnostic significance of the pulse, and distinguished between those nerves that generate sensations and those that move muscles. Herophilus also explored the human genital system, describing the ovaries, the Fallopian tubes, the uterus, and the male prostate gland.

Epicurus. Following the teachings of **Democritus** (see above), Epicurus constructed a philosophical system—Epicureanism—in which the workings of nature were explained not in terms of divine will but in terms of atoms that randomly collide. Death need not be feared, he asserted, because the dispersal of the soul's atoms at death precludes the possibility of an afterlife. The aim of life, he argued, should be a pleasure whose highest expression could be found not through self-indulgent hedonistic excess but through a moderation of desire that is the product of contemplation.

Erasistratus. A physician and medical researcher, Erasistratus described the heart as a four-valve pump that was animated by air coming from the lungs—

though he did not recognize the function of blood in transporting oxygen. Describing the process of digestion, he proposed that food moves through the alimentary canal by a kind of muscular action. Like **Herophilus** (see above), he made a distinction between sensory nerves and motor nerves. After comparing human and animal brains, he theorized that intelligence was a function of the brain's structural complexity.

Euclid. In his enduringly influential multivolume work, *The Elements*, Euclid masterfully synthesized the mathematical and geometrical knowledge of the Greek world up to his time. In separate treatises he applied this knowledge to such fields as optics, harmonics, and astronomy.

Straton (Strato). A believer in experimental research, Straton discovered that falling bodies accelerate, though he lacked the technical means to precisely measure the rate of acceleration. In addition, he discussed at length the nature of a vacuum and the means by which one could be created.

Theophrastus. A colleague of **Aristotle** (see above) and his administrative successor at the Lyceum, Theophrastus became the "father of botany." In addition to his research into plants, he made a scientific study of wind and weather. In psychology, Theophrastus became famous for his satirical profiles of pathological personality types recorded in his book *Characters*.

Third Century BCE

Archimedes. Known for his triumphant declaration "Eureka!" when he unexpectedly found the solution to a formidable puzzle, Archimedes was the most brilliant mathematician of the ancient world and a prolific inventor, much like Leonardo da Vinci. A "one-person Institute for Advanced Study" (to quote mathematician Sherman Stein) and the greatest scientific mind until Newton, Archimedes made astonishing breakthroughs in geometry and physics. He devised an ingenious way to closely calculate the value of pi, grappled with the numerical notion of infinity, and developed a revolutionary method to measure the areas and volumes of a variety of universal shapes whose comparative sizes had confounded all previous mathematicians. He conceived a formula that explained how levers work (famously boasting "Give me a place to stand and I will move the world!") and discovered the principle of specific gravity while taking a bath ("Eureka!"). Archimedes went on to apply principles of mechanics and optics to design devastating engines of war to defend his hometown of Syracuse against Roman attack. He is also said to have constructed a working model of the solar system and a celestial globe.

Eratosthenes. Making projections from his observational data, he was the first scientist to calculate the circumference of Earth and to discover that Earth's axis was tilted relative to the sun. An astronomer, geographer, historian, and literary critic, Eratosthenes was selected by Egypt's king to head the research program at the world-renowned Library of Alexandria.

Sostratus. He designed the famous Pharos, or lighthouse, of Alexandria—one of the original Seven Wonders of the World (not to be confused with the first-century BCE medical researcher also named **Sostratus** [see below]).

Third to Second Centuries BCE

Apollonius of Perge. Apollonius was the last great mathematician of antiquity. His fame rests on his comprehensive study of conic sections, including ellipses, in which he summarized and built upon the work of earlier investigators, such as the fourth-century BCE mathematician Menaechmus. He also applied his mathematical expertise to a theoretical discussion of planetary orbits that would enable astronomers to predict a planet's location.

Diocles. Using his mathematical expertise, Diocles designed a parabolic mirror that could concentrate light and set objects on fire (not to be confused with the fourth-century BCE medical researcher of the same name [see above]).

Philon (Philo) of Byzantium. A mechanical genius, Philon designed a coin-operated vending machine that dispensed holy water and soap. He also employed his talents to invent military catapults and siege equipment. His experiments with air showed that air expands when heated and that air in a closed container can be consumed by flame. Besides inventing and experimenting, Philon authored a (now lost) book on the science of cryptography.

Second Century BCE

Hipparchus. Hipparchus was the ancient world's greatest astronomer. Based on the island of Rhodes, he cataloged 850 stars together with their magnitudes and their precise celestial coordinates, the latter permitting him to predict their future locations in the heavens. Hipparchus also discovered the precession of the equinoxes, and invented devices for celestial navigation. Using the principle of parallax, he calculated the distance from Earth to the moon far more accurately than had his predecessor, **Aristarchus of Samos** (see above).

Seleucus. Based in Babylonia, Seleucus was virtually the only defender of Aristarchus's heliocentric theory of the universe against the more popular view that the sun circled Earth. He also investigated the role of the moon in causing the tides.

Second to First Century BCE

Posidonius. A vastly popular lecturer with wide-ranging intellectual curiosity and knowledge, Posidonius calculated the size of the sun more accurately than had anyone before his time and, in his study of the stars, took atmospheric refraction into account. His underestimation of Earth's size, however, was accepted and mistakenly perpetuated by Ptolemy, ironically inspiring Christopher Columbus to undertake his epoch-making voyage across the Atlantic on the false assumption that he would speedily reach the Orient.

First Century BCE

Asclepiades of Bithynia. A physician, Asclepiades was the first to propose that corpuscles circulated through the human body.

Maria the Jewess. The "founding mother of Western alchemy" and one of the world's first female engineers, this Alexandrian chemist designed an early still and the world's first double boiler for use in her laboratory.

Sosigenes. This Alexandrian astronomer, perhaps at Cleopatra's suggestion, was invited by Julius Caesar in 47 BCE to reform and modernize Rome's antiquated calendar. The result was the Julian calendar, consisting of 3 years of 365 days followed by a single leap year of 366.

Sostratus. An Alexandrian surgeon and distinguished biological researcher (not to be confused with the third-century BCE designer of Alexandria's lighthouse, mentioned above), he ranked next to **Aristotle** in his knowledge of zoology. In the field of medical research, Sostratus's specialty was gynecology.

First Century BCE to First Century CE

Strabo. Strabo authored the most scientifically comprehensive geographical treatise the ancient world ever produced. Remarkably, sixteen of its original seventeen volumes still survive, providing us with an extraordinarily detailed topographical and cultural landscape of the early Roman Empire. Strabo divided the known world into the frigid, temperate, and tropic zones we still refer to today. Though he died in 25 CE, Strabo recognized that Mt. Vesuvius was volcanic in nature, despite the fact that it had not erupted in the memory of humankind. The destruction of Pompeii would not occur for fifty-four more years.

First Century CE

Dioscorides. Attached to Roman armies on the march, he traveled widely throughout the Mediterranean world in search of healing remedies, identifying over a thousand natural drugs. His is the earliest systematic pharmaceutical handbook to survive from the ancient world.

Heron (Hero) of Alexandria. Perhaps the single most outstanding inventor of the Hellenistic age, Heron designed automatic devices astounding in their effects. These included an automated theater, doors that opened and shut automatically, and metal birds that flapped their wings and sang. He demonstrated that air had substance and could be compressed, and he devised an odometer that could function on both land and sea. Though Heron did not recognize its industrial applications in an era when slave labor was abundant, he invented the world's first steam engine, a hollow sphere that spun when water inside turned to steam and exited through vents. His use of lettered diagrams to illustrate mathematical propositions laid the groundwork for the later development of algebra.

Second Century CE

Aelius Aristides. Combined with his hypochondria, his journalistic talents provide us with accounts of the miraculous cures he witnessed at a famous shrine of Asclepius.

Ptolemy (Claudius Ptolemaeus). His astronomical study, the *Mathematike Syntaxis*, or *Almagest*, offered a mathematical model to explain the mechanics of the heavens, but one based incorrectly on a geocentric (rather than a heliocentric) perspective. Building on the work of **Hipparchus** (see above), his catalog of stars listed forty-eight constellations bearing the same names that are still used today. Ptolemy's *Geography* presented in comprehensive form the entire geographical knowledge of his age. His work was to have lasting influence on European thought down to the Renaissance.

Second to Third Century CE

Galen. A court physician in the days of the Roman Empire who had practiced mending the wounds of gladiators, Galen is most famous for his studies of human physiology and comparative anatomy, the latter derived from his experiments on live animals like monkeys. In particular, his research revealed the critical importance of the spinal cord and the deleterious effects of stress on physical well-being.

Third Century CE

Diophantus. Building upon the mathematical studies of **Heron** (see above), Diophantus introduced the equivalent of algebraic equations into a field previously dominated by words and diagrams.

Zosimus. Living in Alexandrian Egypt, Zosimus compiled the first comprehensive encyclopedia of alchemy. Though seemingly more mystical than scientific, the Hellenistic practice of alchemy led to important discoveries in chemistry, including the development of equipment still used in research laboratories today.

Fourth to Fifth Century CE

Hypatia. Daughter of the Alexandrian mathematician Theon, Hypatia excelled in mathematics and edited the works of such earlier figures as **Ptolemy** (see above). A charismatic proponent of Neoplatonism, she is said to have been brutally murdered by a mob of fanatical Christians.

NOTES

CHAPTER 1

1. From an address presented at the inauguration of the Faculty of Science, University of Lille, December 7, 1854. Cited in *The Concise Oxford Dictionary of Quotations*, 2nd ed. (New York: Oxford University Press, 1981), p. 181.

2. For a comprehensive discussion of the ideals of Greek civilization and their relevance to our lives, see Stephen Bertman, *The Eight Pillars of Greek Wisdom* (New York: Barnes & Noble, 2007).

3. Mary McCarthy, "The Fact in Fiction," in *On the Contrary* (New York: Farrar, Straus, and Cudahy, 1961).

CHAPTER 2

1. Herodotus, *History*, 2.5.

2. Ibid., 2.86–90.

3. For a detailed survey of Mesopotamian civilization, see Stephen Bertman, *Handbook to Life in Ancient Mesopotamia* (New York: Facts on File, 2003; Oxford University Press, 2005).

4. *History*, 1.197.

5. Exod., 14:10–15:21.

6. 2 Kings, 20:8–11.

7. Josh., 10:12–14.

8. Gen., 2:15–17.

9. Ibid., 3:1–24.

10. Prov., 9:10; cf. 1:7.

11. Job, 38:4.

12. Ibid., 40:4–5.

13. Ibid., 42:2–3, 6.

14. Eccles., 1:2.

15. Ibid., 1:18, 12:12.

16. Ibid., 8:17.

17. Ibid., 2:16.

CHAPTER 3

1. Vitruvius, *On Architecture*, 6.1.
2. Aristoxenus, *Fragm.* 23 (Wehrli).

CHAPTER 4

1. Homer, *Iliad*, 17.645–47.
2. Homer, *Odyssey*, 9.116–566.
3. Ibid., 1.3.
4. Aeschylus, *Agamemnon*, 1–39.
5. Heron of Alexandria, *Catoptrics*, 1.
6. From the Moorish-geographer Idrisi's twelfth-century work *Description of Africa and Spain*.
7. John Zonaras, *Epitome of the Histories*, 9.4.
8. John Tzetzes, *Book of Histories*, 2.118–28.
9. Aristophanes, *The Clouds*, 763–73.
10. Euclid, *Optics*, Definitions and Propositions I–VIII, XLV, and XLVIII (Heiberg).

CHAPTER 5

1. See, for example, *Iliad*, 1.57–305, 2.207–399, and *Odyssey*, 1.6–259, 24.413–66.
2. Nico F. Declercq and Cindy S. A. Dekeyser, "Acoustic Diffraction Effects at the Hellenistic Amphitheater of Epidaurus: Seat Rows Responsible for the Marvelous Acoustics," *Journal of the Acoustical Society of America* 121.4 (April 2007): 2011–22.
3. Vitruvius, *On Architecture*, 5.3.8.

CHAPTER 6

1. Aristotle, *Nicomachean Ethics*, 1181.12–15.
2. Homer, *Odyssey*, 11.595–600.
3. John Conduitt, *Keynes Ms.*, 130.4. Source: The Newtonproject.
4. Aristotle, *Physics*, 7.5.

5. Ibid., 4.8.

6. Recorded in Simplicius's commentary on Aristotle's *Physics*, 1110.5.

7. See "Aristotle," *Mechanics*, 4, 21.

8. See Plutarch's *Parallel Lives*, "Life of Marcellus," 14.3–7, and, for a more general Greek attitude toward such matters, Aristotle, *Politics*, 7.6.3ff. and especially 15–16.

9. Livy, *History of Rome*, 24.34.10–11.

10. Plutarch, "Life of Marcellus," 15.3.

11. Heron of Alexandria, preface to *Artillery Manual*, pp. 71–73 (Wescher).

12. Diodorus Siculus, *Library*, 4.76.2.

13. Reported by Aristotle, *On the Soul*, 1.3.

14. Apollonius Rhodius, *Argonautica*, 4:1638–48.

15. At least according to the twelfth-century CE Byzantine writer Eustathius (*On the Odyssey*, 20.302).

16. *Iliad*, 18:368–79, 417–20.

17. Aristotle, *On the Movements of Animals*, 7–13, *On the Generation of Animals*, 2.1, 2.5, and *Mechanics*, 1.

18. Aristotle, *On Dreams*, 3.

19. Heron of Alexandria, *The Automatic Theater*, 4.1–3, 14.1–2.

20. Heron of Alexandria, introduction to *Pneumatics*, rev. ed., trans. Bennett Woodcroft.

21. Ibid., 57.

22. T. E. Rihll, *Greek Science* (New York: Oxford University Press, 1999), p. 35.

23. Heron, *Pneumatics*, 50, trans. Bennett Woodcroft.

CHAPTER 7

1. Plato, *Apology*, 38A.

2. Issa Tapsoba et al., "Finding Out Egyptian Gods' Secret Using Analytical Chemistry: Biomedical Properties of Egyptian Black Makeup Revealed by Ampherometry at Single Cells," *Analytical Chemistry* 82, no. 2 (December 2009): 457–60.

3. Vitruvius, *On Architecture*, 9.9–12.

CHAPTER 8

1. Herodotus, *History*, 4.152.

2. Aristides, 36.85–95.

3. Arrian, *Indica*, 43.11–12.

4. Duane Roller, *Through the Pillars of Herakles: Greco-Roman Exploration of the Atlantic* (New York: Routledge, 2006), p. xi.

5. Herodotus, *History*, 4.44.

6. Strabo, *Geography*, 1.1.1.

7. See, for example, *Iliad*, 18.399, pp. 607–608, and *Odyssey*, 11.11–19.

8. Herodotus, *History*, 4.36.

9. For the *Iliad*, see Cedric H. Whitman, *Homer and the Heroic Tradition* (Cambridge, MA: Harvard University Press, 1958); for the *Odyssey*, see Stephen Bertman, *Analogy and Contrast as Elements of Symmetrical Design in the Structure of the* Odyssey (PhD diss., Columbia University, 1965) and "Structural Symmetry at the End of the Odyssey," *Greek, Roman and Byzantine Studies* 9 (1968): 115–23.

10. See John L. Myres, *Herodotus, Father of History* (Oxford: Clarendon Press, 1953).

11. Hesiod, *Theogony*, 126–32.

CHAPTER 9

1. Homer, *Odyssey*, 10.47–49.

2. Hesiod, *Works and Days*, 505–18, abridged.

CHAPTER 10

1. Homer, *Iliad*, 18.483–89.

2. Homer, *Odyssey*, 5.269–81.

3. Hesiod, *Works and Days*, 597–617.

4. Xenophanes, *Fragm.* 13 (Diehl).

5. Plato, *Theaetetus*, 174A.

6. Aristotle, *Politics*, 1.11.

7. Herodotus, *History*, 1.74.2.

8. See the discussion in his *Timaeus*.

9. *Republic*, 7.527D–530C.

10. Cicero, *De Re Publica*, 1:21–2.

11. G. J. Toomer, "Aratus," in *The Oxford Classical Dictionary*, 3rd ed. (New York: Oxford University Press, 1966), pp. 136–37.

12. Guillaume Bigourdan, *L'Astronomie: Evolution des Idées et des Méthodes* (Paris: Flammarion, 1911), p. 279, cited in Arnold Reymond, *History of the Sciences in Greco-Roman Antiquity*, trans. Ruth Gheury de Bray (London: Methuen, 1927), p. 87.

13. For a discussion of the astronomical significance of the statue, see Kenneth Chang, "Visions of Ancient Night Sky Were Hiding in Plain Sight for Centuries," *New York Times*, January 18, 2005.

14. For an up-to-date discussion of the Antikythera mechanism in the light of ongoing research, see Jo Marchant, *Decoding the Heavens: A 2,000-Year-Old Computer and the Century-Long Search to Discover Its Secrets* (New York: Da Capo, 2009).

15. *Almagest*, 1.1.

16. G. E. R. Lloyd, *Methods and Problems in Greek Science: Selected Papers* (Cambridge: Cambridge University Press, 1991), p. 162.

CHAPTER 11

1. Homer, *Iliad*, 11.489–91.

2. Ibid., 17.426ff., 19.392ff.

3. Homer, *Odyssey*, 17.291ff.

4. Gen., 1:11–25.

5. Ibid., 1:26–31.

6. Hesiod, *Works and Days*, 106ff.

7. Apollodorus, *The Library*, 1.7.2.

8. Ovid, *Metamorphoses*, 1.253–415.

9. Gen., 7:1–8:19.

10. Herodotus, *History*, 1.1.

11. Galen, *On the Doctrines of Hippocrates and Plato*, 5.3.

12. Vitruvius, *On Architecture*, 3.1.1–2.

13. Ibid., 3.1.3.

14. Aristotle, *On the Soul*, 3.3, *History of Animals*, 1.1.16, and *On the Parts of Animals*, 1.1.

15. Aristotle, *On the Parts of Animals*, 1.5.

16. Ibid.

17. Aristotle, *History of Animals*, 6.3.

18. Ibid., 9.40.

19. For the concept of such a scale or ladder, see Loren Eiseley, *Darwin's Century:*

Evolution and the Men Who Discovered It (New York: Barnes & Noble, 2009 [1958]), pp. 4–9.

20. Aristotle, *On the Parts of Animals*, 2.10.

21. Ibid., 3.4.

22. Aristotle, *History of Animals*, 2.3, *On the Parts of Animals*, 3.1.

23. Theophrastus, *History of Plants*, 1.1, trans. A. G. Morton.

24. Ibid., 1.7.1–2.

25. Ibid., 1.13.2.

26. Ibid., 2.2.11–12.

27. Erasistratus, *On Paralysis*, quoted by Galen in *On Habits*, 1.

28. Galen, *On the Use of Parts*, 3.20, trans. G. E. R. Lloyd.

CHAPTER 12

1. Homer, *Iliad*, 4.480–82.

2. Ibid., 4.521–26.

3. Ibid., 11.828–36.

4. Homer, *Odyssey*, 4.220–21.

5. Ibid., 4.228–32.

6. Ibid., 9.82–104.

7. Hippocrates, "Prognostic," *Hippocratic Collection*, 1.

8. G. E. R. Lloyd, *Greek Science after Aristotle* (New York: Norton, 1973), p. 137.

9. Thucydides, *The Peloponnesian War*, 2.47–54.

10. Ibid., 2.49.2–8.

11. Ibid., 2.52.2–53.

12. Ibid., 1.22.

13. Plato, *Phaedo*, 116C–118.

CHAPTER 13

1. Homer, *Iliad*, 1.1–5.

2. Ibid., 1.362–63.

3. Aristotle, *Poetics*, 1449B.

4. Sigmund Freud, *Delusion and Dream: An Interpretation in the Light of Psychoanalysis of* Gradiva, *a Novel, by Wilhelm Jensen*, trans. Helen M. Downey (London: George Allen & Unwin, 1921), p. 113.

5. William Shakespeare, *Julius Caesar*, act 1, scene 2.

6. Ibycus, 1, in *Erotic Love Poems of Greece and Rome*, transl. Stephen Bertman (New York: New American Library, 2005), p. 40.

7. Philodemus, 11.30, ibid., pp. 58–59.

8. *The Alexandrian Erotic Fragment*, ibid., pp. 48–49.

9. Aristotle, *On the Soul*, 3.3, and *Nicomachean Ethics*, 1.7–9.

10. Aristotle, *On Dreams*.

11. Aristotle, *Rhetoric*, 2.12–14.

12. Plato, *Republic*, 514A–521B.

13. Ibid., 434D–441C.

CHAPTER 14

1. Plutarch, *Parallel Lives*, "Life of Cato the Elder," 20–23.

2. Horace, Epistles, 2.1.156.

3. Virgil, *Aeneid*, 6.847–63.

4. Frontinus, *The Aqueducts of Rome*, 16.

5. Plutarch, *Parallel Lives*, "Life of Marcellus," 1.

6. Ibid., 19.4–6.

7. Cicero, *Tusculan Disputations*, 1.2.

8. Ibid., 5.64–66.

9. Vitruvius, *On Architecture*, 1.1.3.

10. Ibid., 1.1.4–10.

11. *Aetna*, 224–51, abridged.

12. Seneca, *Natural Questions*, 7.25.4–7.

13. Lynn Thorndike, *A History of Magic and Experimental Science during the First Thirteen Centuries of Our Era*, vol. 1 (New York: Columbia University Press, 1923), pp. 42–43.

14. Ibid., pp. 54–56.

15. Ian Gray Kidd, "Posidonius," in *The Oxford Classical Dictionary*, 3rd ed. (New York: Oxford University Press, 1996), p. 1232.

16. Seneca, *Natural Questions*, 7.32.4.

17. Pliny, *Natural History*, 2.15.118.

Chapter 15

1. Ammianus Marcellinus, *History*, 14.6.18.

2. St. Jerome, *Letter 127.12 ("To Principia," 412 CE)*.

3. Thomas Hobbes, *Leviathan*, chap. 13.

4. Gilbert Highet, *The Classical Tradition: Greek and Roman Influence on Western Literature* (New York: Oxford University Press, 1976), p. 8.

5. Ibid., pp. 13–14.

6. St. Augustine, *Enchiridion*, 9.

7. Marshall Clagett, *Greek Science in Antiquity* (Mineola, NY: Dover, 2001 [1955]), p. 130.

8. Thomas Goldstein, *Dawn of Modern Science: From the Ancient Greeks to the Renaissance* (New York: Da Capo, 1995), p. 108.

9. Cited by Ray Spangenburg and Diane K. Moser, *The History of Science: From the Ancient Greeks to the Scientific Revolution* (New York: Facts on File, 1993), p. 26.

10. Reviel Netz and William Noel, *The Archimedes Codex: How a Medieval Prayer Book Is Revealing the True Genius of Antiquity's Greatest Scientist* (New York: Da Capo, 2007), p. 1.

11. For further information about the Archimedes Palimpsest and for images of its pages, go to http://www.archimedespalimpsest.org.

12. Gilbert Highet, *The Classical Tradition*, p. 15, citing J. A. Symonds.

13. See Jacob Bronowski, "Leonardo da Vinci," in *The Horizon Book of the Renaissance* (New York: American Heritage, 1961), p. 189.

Chapter 16

1. Stephen Bertman, *Doorways through Time: The Romance of Archaeology* (Los Angeles: Tarcher, 1987), pp. 195–96.

Chapter 17

1. Diodorus, *History*, 2.47.

2. Julius Caesar, *Commentaries on the Gallic War*, 6.14.

3. William Stukeley, *Stonehenge: A Temple Restor'd to the British Druids* (London: W. Innys and R. Manby), pp. 1740–43.

4. See Norman Lockyer, *Stonehenge and Other British Stone Monuments Astronomically Considered* (London: Macmillan, 1909).

5. See Gerald S. Hawkins, *Stonehenge Decoded* (New York: Dorset, 1965).

CHAPTER 18

1. Dick Teresi, *Lost Discoveries: The Ancient Roots of Modern Science—From the Baby-lonians to the Maya* (New York: Simon & Schuster, 2002), p. 146.

2. For a comparative study of ancient Greek and Chinese cultural and scientific attitudes, see G. E. R. Lloyd, *Ancient Worlds, Modern Reflections: Philosophical Perspectives on Greek and Chinese Science and Culture* (Oxford: Oxford University Press, 2004).

3. Stephen Bertman, *Doorways through Time: The Romance of Archaeology* (Los Angeles: Tarcher, 1987), p. 187.

EPILOGUE

1. For further perspective on this theme, see G. E. R. Lloyd, *Methods and Problems in Greek Science: Selected Papers*, esp. chaps. 15 and 18 (Cambridge: Cambridge University Press, 1991), and T. E. Rihll, *Greek Science*, chap. 1 (Oxford: Oxford University Press, 1999).

2. Sophocles, *Antigone*, 335–64.

RECOMMENDED READINGS

SOURCES

Cohen, Morris R., and I. E. Drabkin. *A Source Book in Greek Science.* Cambridge, MA: Harvard University Press, 1966.

Freeman, Kathleen. *Ancilla to the Pre-Socratic Philosophers: A Complete Translation of the Fragments in Diels'* Fragmente der Vorsokratiker. Forgotten Books, 2008; Oxford: Blackwell, 1952.

Guthrie, Kenneth Sylvan, and David Fideler, eds. *The Pythagorean Sourcebook and Library: An Anthology of Ancient Writings Which Relate to Pythagoras and Pythagorean Philosophy.* Grand Rapids, MI: Phanes Press, 1987.

Heath, Thomas L., ed. *The Thirteen Books of Euclid's Elements.* 3 vols. Mineola, NY: Dover, 1956 (1908).

———. *The Works of Archimedes.* Mineola, NY: Dover, 2002 (1897).

Humphrey, John W. *Greek and Roman Technology: A Sourcebook.* New York: Routledge, 1997.

Irby-Massie, Georgia L., and Paul T. Keyser. *Greek Science of the Hellenistic Era: A Sourcebook.* New York: Routledge, 2002.

Marsden, E. W. *Greek and Roman Artillery: Technical Treatises.* London: Oxford University Press, 1971.

Netz, Reviel, ed. and trans. *The Works of Archimedes: Translation and Commentary.* New York: Cambridge University Press, 2009.

Rorres, Chris. *Archimedes Home Page.* http://www.mcs.drexe.edu/~crorres/Archimedes.

Singer, Peter N., trans. *Galen: Selected Works.* New York: Oxford University Press, 2002.

Thomas, Ivor, trans. *Selections Illustrating the History of Greek Mathematics.* 2 vols. Cambridge, MA: Harvard University Press, rev. ed., 1991 and 1993.

The Loeb Classical Library published by Harvard University Press (http://www.hup .harvard.edu/loeb) presents the works of classical authors (including Aristotle and many others) in editions with the original Greek or Latin text and an English translation on opposite pages. Many original and translated classical texts, together with supplemental information, can also be found in the Perseus Digital Library (http://www.perseus.tufts.edu).

SURVEYS OF ANCIENT SCIENCE

Asimov, Isaac. *Asimov's Biographical Encyclopedia of Science and Technology.* 2nd ed. Garden City, NY: Doubleday, 1982.

Clagett, Marshall. *Greek Science in Antiquity.* New York: Abelard-Schuman, 1955; Mineola, NY: Dover, 2001.

Cuomo, Serafina. *Technology and Culture in Greek and Roman Antiquity.* New York: Cambridge University Press, 2007.

De Camp, L. Sprague. *The Ancient Engineers.* New York: Ballantine, 1963.

Farrington, Benjamin. *Greek Science: Its Meaning for Us.* Nottingham: Spokesman, 2000 (1953).

_____. *Science and Politics in the Ancient World.* 2nd ed. New York: Barnes & Noble, 1966.

_____. *Science in Antiquity.* 2nd ed. London & New York: Oxford University Press, 1969.

Gillespie, Charles Coulston, ed. *Dictionary of Scientific Biography.* 16 vols. New York: Scribner's, 1981.

Goldstein, Thomas. *Dawn of Modern Science: From the Ancient Greeks to the Renaissance.* Foreword by Isaac Asimov. New York: Da Capo, 1995.

Hodges, Henry. *Technology in the Ancient World.* New York: Barnes & Noble, 1970.

Keyser, Paul T., and Georgia Irby-Massie. *Encyclopedia of Ancient Natural Scientists.* New York: Routledge, 2008.

Lang, Phillipa. "Science, Greece" and "Science, Rome" in *Encyclopedia of Society and Culture in the Ancient World,* Vols. 4:937–41 and 941–51, edited by Peter Bogucki (New York: Facts on File, 2008).

Lindberg, David C. *The Beginnings of Western Science: The European Scientific Tradition in Philosophical, Religious, and Institutional Context, Prehistory to AD 1450.* 2nd ed. Chicago: University of Chicago Press, 2007.

Lloyd, G. E. R. *Early Greek Science: Thales to Aristotle.* New York: Norton, 1970.

_____. *Greek Science after Aristotle.* New York: Norton, 1973.

_____. *Polarity and Analogy: Two Types of Argumentation in Early Greek Thought.* Cambridge: Cambridge University Press, 1966.

Nardo, Don. *Greek and Roman Science.* San Diego, CA: Lucent, 1998.

Neugebauer, Otto. *The Exact Sciences in Antiquity.* 2nd ed. Providence, RI: Brown University Press, 1957.

Oleson, John Peter, ed. *The Oxford Handbook of Engineering and Technology in the Classical World.* New York: Oxford University Press, 2008.

Reymond, Arnold. *History of the Sciences in Greco-Roman Antiquity.* Trans. Ruth Gheury de Bray. London: Methuen, 1927.

Rihll, T. E. *Greek Science (Greece & Rome,* "New Surveys in the Classics, No. 29"). Oxford: Oxford University Press (for the Classical Association), 1999.

Sarton, George. *Ancient Science through the Golden Age of Greece.* Mineola, NY: Dover 1993 (1952).

Spangenburg, Ray, and Diane K. Moser. *The History of Science from the Ancient Greeks to the Scientific Revolution.* New York: Facts on File, 1993.

Stahl, William H. *Roman Science: Origins, Development, and Influence to the Later Middle Ages.* Westport, CT: Greenwood, 1978 (1962).

Temple, Robert. *The Genius of China: 3,000 Years of Science, Discovery & Invention.* Foreword by Joseph Needham. Rochester, VT: Inner Tradition, 2007 (1985).

SURVEYS OF ANCIENT GREEK CIVILIZATION

Bertman, Stephen. *The Eight Pillars of Greek Wisdom.* New York: Fall River/Barnes & Noble, 2007.

Brunschwig, Jacques, and G. E. R. Lloyd. *Greek Thought: A Guide to Classical Knowledge.* Cambridge, MA: Harvard University Press, 2000.

Cahill, Thomas. *Sailing the Wine-Dark Sea: Why the Greeks Matter.* New York: Doubleday, 2003.

Durant, Will. *The Life of Greece.* Vol. 2 of *The Story of Civilization.* New York: Simon & Schuster, 1939.

Finley, Sir Moses I. *The Legacy of Greece: A New Appraisal.* New York: Oxford University Press, 1984.

Freeman, Charles. *The Greek Achievement: The Foundation of the Western World.* New York: Penguin, 2000.

Jaeger, Werner. *Paideia: The Ideals of Greek Culture.* 3 vols. New York: Oxford University Press, 1944.

Kitto, H. D. F. *The Greeks.* New York: Penguin, 1957.

Livingstone, R. W., ed. *The Legacy of Greece.* Oxford: Clarendon Press, 1921.

SPECIAL STUDIES

Africa, Thomas W. *Science and the State in Greece and Rome.* New York: John Wiley, 1968.

Aveni, Anthony F. *Ancient Astronomers.* Washington, DC: Smithsonian Press, 1993.

_____. "Apocalypse Soon? What the Maya Calendar Really Tells Us about 2012 and the End of Time." *Archaeology* 62, no. 6 (November–December 2009): 30–35.

_____. *Conversing with the Planets: How Science and Myth Invented the Cosmos.* New York: Time Books, 1992.

_____. *Skywatchers.* Rev. and enlarged ed. of *Skywatchers of Ancient Mexico,* 1980. Austin: University of Texas Press, 2001.

_____. *Stairways to the Stars: Skywatching in Three Great Ancient Cultures.* New York: Wiley, 1997.

Barker, Andrew. *The Science of Harmonics in Classical Greece.* Cambridge: Cambridge University Press, 2007.

Barrow, R. H. *The Romans.* New York: Penguin, 1949.

Barton, Tamsyn. *Ancient Astrology.* New York: Routledge, 1994.

Beagon, Mary. *Roman Nature: The Thought of Pliny the Elder.* New York: Oxford University Press, 1992.

Beck, Roger. *A Brief History of Ancient Astrology.* Hoboken, NJ: Wiley-Blackwell, 2007.

Berryman, Sylvia. *The Mechanical Hypothesis in Ancient Greek Natural Philosophy.* New York: Cambridge University Press, 2009.

Bertman, Stephen. *Doorways through Time: The Romance of Archaeology.* Los Angeles: Tarcher, 1986.

Bertman, Stephen, and Lois Parker, eds. *The Healing Power of Ancient Literature.* Newcastle upon Tyne: Cambridge Scholars Publishing, 2009.

Bowen, Alan C., ed. *Science and Philosophy in Classical Greece.* New York: Garland, 1991.

Boyer, Carl B. *A History of Mathematics.* 2nd ed., edited by Uta C. Merzbach. New York: Wiley, 1989.

Brownworth, Lars. *Lost to the West: The Forgotten Byzantine Empire That Rescued Western Civilization.* New York: Crown, 2009.

Buxton, Richard, ed. *From Myth to Reason? Studies in the Development of Greek Thought.* New York: Oxford University Press, 1999.

Cary, Max, and E. H. Warmington. *The Ancient Explorers.* New York: Penguin, 1963.

Casson, Lionel. *The Ancient Mariners.* 2nd ed. Princeton, NJ: Princeton University Press, 1991.

_____. *Libraries in the Ancient World.* New Haven, CT: Yale University Press, 2001.

_____. *Ships and Seamanship in the Ancient World.* Baltimore, MD: Johns Hopkins University Press, 1995 (1971).

_____. *Travel in the Ancient World.* Baltimore, MD: Johns Hopkins University Press, 1994.

Catherwood, Frederick. *Views of Ancient Monuments in Central America, Chiapas, and Yucatan.* New York: Bartlett and Welford, 1844.

Chippindale, Christopher. "Stonehenge Astronomy: Anatomy of a Modern Myth." *Archaeology* 39, no. 1 (January–February 1986): 47–52.

_____. *Stonehenge Complete.* 3rd ed. New York: Thames & Hudson, 2004.

Clagett, Marshall. *Archimedes in the Middle Ages.* 5 vols. Madison: University of Wisconsin Press; Philadelphia: American Philosophical Society, 1964–1984.

Coe, Michael D. *Mexico: From the Olmecs to the Aztecs.* New York: Thames & Hudson, 1996.

Conrad, Lawrence I., et al. *The Western Medical Tradition.* Cambridge: Cambridge University Press, 1995.

Corben, Herbert C. *The Struggle to Understand: A History of Human Wonder and Discovery.* Amherst, NY: Prometheus, 1991.

Crowe, Michael J. *Theories of the World from Antiquity to the Copernican Revolution.* Mineola, NY: Dover, 1990.

Cunliffe, Barry. *The Extraordinary Voyage of Pytheas the Greek.* Rev. ed. New York: Walker, 2001.

Cuomo, Serafina. *Ancient Mathematics.* New York: Routledge, 2001.

Dawson, Ian. *Greek and Roman Medicine.* New York: Enchanted Lion, 2005.

Dijksterhuis, E. J. *Archimedes, with a New Biographic Essay by Wilbur R. Knorr.* Princeton: Princeton University Press, 1987 (1938).

Dodds, E. R. *The Greeks and the Irrational.* Berkeley: University of California Press, 1951.

Doody, Aude. *Pliny's Encyclopedia: The Reception of the Natural History.* Cambridge: Cambridge University Press, 2010.

Durant, Will. *The Story of Philosophy: The Lives and Opinions of the Greater Philosophers.* Chaps. 1 ("Plato") and 2 ("Aristotle and Greek Science"). New York: Simon & Schuster, 1926.

Dzielska, Maria. *Hypatia of Alexandria.* Cambridge, MA: Harvard University Press, 1995.

Eiseley, Loren. *Darwin's Century: Evolution and the Men Who Discovered It.* Foreword by Stephen Bertman. New York: Barnes & Noble, 2009 (1958).

Ellul, Jacques. *The Technological Society.* New York: Knopf, 1964.

Evans, James. *The History and Practice of Ancient Astronomy.* New York: Oxford University Press, 1998.

Ferguson, Kitty. *The Music of Pythagoras: How an Ancient Brotherhood Cracked the Code of the Universe and Lit the Path from Antiquity to Outer Space.* New York: Walker, 2008.

FitzGerald, Patrick. *Ancient China.* Oxford: Elsevier-Phaidon, 1978.

Forbes, R. J. *Studies in Ancient Technology.* 6 vols. Leiden: Brill, 1963.

Freely, John. *Aladdin's Lamp: How Greek Science Came to Europe through the Islamic World.* New York: Knopf, 2009.

Freeth, Tony, et al. "Calendars with Olympiad Display and Eclipse Prediction on the Antikythera Mechanism." *Nature* 454 (July 31, 2008): 614–17.

———. "Decoding the Ancient Greek Astronomical Calendar Known as the Antikythera Mechanism." *Nature* 444 (November 30, 2006): 587–91.

French, Roger. *Ancient Natural History: Histories of Nature.* New York: Routledge, 1994.

Goodyear, William Henry. *Greek Refinements: Studies in Temperamental Architecture.* New Haven, CT: Yale University Press, 1912.

Hagel, Stefan. *Ancient Greek Music: A New Technical History.* New York: Cambridge University Press, 2009.

Hawkins, Gerald S. *Stonehenge Decoded.* New York: Dorset, 1965.

Heath, Thomas L. *Archimedes.* New York: Macmillan, 1920.

———. *Greek Astronomy.* Mineola, NY: Dover 1991 (1932).

———. *A History of Greek Mathematics.* Oxford: Oxford University Press, 1921.

Hetherington, Norriss S. *Ancient Astronomy and Civilization.* Illustrations by A. V. Mann. Tucson, AZ: Pachart Publishing House, 1987.

Hibbert, Christopher. *The Emperors of China.* Chicago: Stonehenge Press, 1981.

Highet, Gilbert. *The Classical Tradition: Greek and Roman Influence on Western Literature.* New York: Oxford University Press, 1976.

Hirshfeld, Alan. *Eureka Man: The Life and Legacy of Archimedes.* New York: Walker, 2009.

Hodson, F. R., ed. *The Place of Astronomy in the Ancient World.* London: Oxford University Press, 1974.

Huffman, Carl A. *Philolaus of Croton: Pythagorean and Presocratic.* Cambridge: Cambridge University Press, 1993.

"Infinite Secrets: The Genius of Archimedes." *NOVA* (BBC/WGBH Boston), 2003.

Jaeger, Mary. *Archimedes and the Roman Imagination.* Ann Arbor: University of Michigan Press, 2008.

James, Peter, and Nick Thorpe. *Ancient Inventions.* New York: Ballantine, 2006 (1994).

Kahn, Charles H. *Pythagoras and the Pythagoreans: A Brief History.* Indianapolis, IN: Hackett, 2001.

Keyser, Paul T. "Alchemy in the Ancient World: From Science to Magic." *Illinois Classical Studies* 15 (1990): 353–78.

———. "Greco-Roman Alchemy and Coins of Imitation Silver." *American Journal of Numismatics* 7/8 (1995/6): 209–234, plates 28–32.

———. "A New Look at Heron's Steam Engine." *Archive for History of Exact Sciences* 44 (1992): 107–24.

———. "The Purpose of the Parthian Galvanic Cells: A First-Century A.D. Electric Battery Used for Analgesi." *Journal of Near Eastern Studies* 52 (1993): 81–98.

King, Helen. *Greek and Roman Medicine.* London: Duckworth, 2001.

Krebs, Robert E. and Carolyn A. *Groundbreaking Scientific Experiments, Inventions, and Discoveries of the Ancient World.* Westport, CT: Greenwood, 2003.

Krupp, E. C. *Echoes of the Ancient Skies: The Astronomy of Lost Civilizations.* New York: Oxford University Press, 1983.

Landels, J. G. *Engineering in the Ancient World.* Rev. ed. Berkeley and Los Angeles: University of California Press, 2000.

Lang, Phillipa, ed. *Reinventions: Essays on Hellenistic and Early Roman Science.* Kelowna, BC: Academic Printing and Publishing, 2004 (*Apeiron*, 37.4 [Dec. 2004]).

Lawrence, A. W. *Greek Architecture.* Chap. 15, "Niceties of Doric Design." Baltimore, MD: Penguin, 1957.

Levin, Flora R. *Greek Reflections on the Nature of Music.* Cambridge: Cambridge University Press, 2009.

Lindberg, David C., ed. *Science in the Middle Ages.* Chicago: University of Chicago Press, 1978.

Lindsay, Jack. *Blast Power and Ballistics: Concepts of Force and Energy in the Ancient World.* New York: Barnes & Noble, 1974, 2009.

_____. *The Origins of Alchemy in Graeco-Roman Egypt.* New York: Barnes & Noble, 1970.

Lloyd, G. E. R. *The Ambitions of Curiosity: Understanding the World in Ancient Greece and China.* New York: Cambridge University Press, 2002.

_____. *Ancient Worlds, Modern Reflections: Philosophical Perspectives on Greek and Chinese Science and Culture.* Oxford: Oxford University Press, 2004.

_____. *The Revolutions of Wisdom: Studies in the Claims and Practice of Ancient Greek Science.* Berkeley: University of California Press, 1987.

_____. *Science, Folklore and Ideology.* London: Duckworth, 1999 (1983).

Majno, Guido. *The Healing Hand: Man and Wound in the Ancient World.* Cambridge, MA: Harvard University Press, 1975.

Maor, Eli. *The Pythagorean Theorem: A 4,000-Year History.* Princeton, NJ: Princeton University Press, 2007.

Marchant, Jo. *Decoding the Heavens: Solving the Mystery of the World's First Computer.* New York: Da Capo, 2009.

Marsden, E. W. *Greek and Roman Artillery: Historical Development.* Oxford: Clarendon Press, 1969.

Mattern, Susan P. *Galen and the Rhetoric of Healing.* Baltimore: Johns Hopkins University Press, 2008.

Murphy, S. E. "Heron of Alexandria's 'On Automata-Making.'" *History of Technology* 17 (1994): 1–44.

Needham, Joseph. *Science and Civilization in China.* 7 vols. Cambridge: Cambridge University Press, 2009 (1954–2004).

Netz, Reviel. *Ludic Proof: Greek Mathematics and the Alexandrian Aesthetic.* New York: Cambridge University Press, 2009.

_____. *The Shaping of Deduction in Greek Mathematics: A Study in Cognitive History.* New York: Cambridge University Press, 1999.

_____. *The Transformation of Mathematics in the Early Mediterranean World: From Problems to Equations.* New York: Cambridge University Press, 2004.

Netz, Reviel, and William Noel. *The Archimedes Codex: How a Medieval Prayer Book Is Revealing the True Genius of Antiquity's Greatest Scientist.* New York: Da Capo, 2007.

Neugebauer, O. *History of Ancient Mathematical Astronomy.* 3 vols. Berlin: Springer, 1975.

Nicastro, Nicholas. *Circumference: Eratosthenes and the Ancient Quest to Measure the Globe.* New York: St. Martin's Press, 2008.

Nutton, Vivian. *Ancient Medicine.* New York: Routledge, 2004.

O'Connor, J. J., and E. F. Robertson. "How Do We Know about Greek Mathematics?" http://www.gap.dcs.st-and.ac.uk/~history/HistTopics/Greek_sources_1.html.

O'Leary, DeLacy. *How Greek Science Passed to the Arabs.* London: Routledge and Kegan Paul, 1949.

Patai, Raphael. "Maria the Jewess: Founding Mother of Alchemy." *Ambix* 29.3 (1982): 177–97.

Pearsall, Ronald. *The Alchemists.* London: Weidenfeld and Nicolson, 1976.

Penrose, F. C. *An Investigation of the Principles of Athenian Architecture; or The Results of a Recent Survey Conducted Chiefly with Reference to the Optical Refinements Exhibited in the Construction of the Ancient Buildings at Athens.* London: Nicol (for the Society of Dilettanti), 1851.

Posamentier, Alfred S. *The Pythagorean Theorem: The Story of Its Beauty and Power.* Amherst, NY: Prometheus, 2010.

Price, Derek J. de Solla. "An Ancient Greek Computer." *Scientific American* 200, no. 6 (June 1959): 60–67.

Rice, Rob S. "The Antikythera Mechanism: Physical and Intellectual Salvage from the First Century BC." *USNA Eleventh Naval History Symposium,* 1995. http://ccat.sas.upenn.edu/rrice/usna_pap_fn.html.

Rocca, Julius. *Reading Ancient Medical Writers.* New York: Routledge, 2010.

Roccatoagliata, Giuseppe. *A History of Ancient Psychiatry.* Westport, CT: Greenwood, 1986.

Rogers, John H. "Origins of the Ancient Constellations." *Journal of the British Astronomical Association* 108 (1998): part I "The Mesopotamian Traditions," 9–28; part II "The Mediterranean Traditions," 79–89.

Roller, Duane W. *Eratosthenes' "Geography."* Princeton, NJ: Princeton University Press, 2009.

_____. *Through the Pillars of Herakles: Greco-Roman Exploration of the Atlantic.* New York: Routledge, 2006.

Romm, James S. *The Edges of the Earth in Ancient Thought: Geography, Exploration, and Fiction.* Princeton, NJ: Princeton University Press, 1992.

Rosen, William. *The Most Powerful Idea in the World: A Story of Steam, Industry, and Invention.* New York: Random House, 2010.

Rudman, Peter S. *The Babylonian Theorem: The Mathematical Journey of Pythagoras and Euclid.* Amherst, NY: Prometheus, 2009.

Samburksy, Shmuel. *Physical Thought from the Presocratics to the Quantum Physicists: An Anthology.* New York: Pica, 1975.

————. *The Physical World of the Greeks.* New York: Collier, 1962.

Santandar, Mariano. "The Chinese South-Seeking Chariot." *American Journal of Physics* 60, no. 9 (September 1992): 782–87.

Sarton, George. *Ancient Science and Modern Civilization.* Lincoln: University of Nebraska Press, 1954.

Simms, D. L. "Archimedes and the Burning Mirrors of Syracuse." *Technology and Culture* 18.1 (1977): 1–24.

Stephens, John Lloyd. *Incidents of Travel in Central America, Chiapas, and Yucatan.* London: John Murray, 1842.

Stein, Sherman. *Archimedes: What Did He Do Besides Cry Eureka?* Washington, DC: Mathematical Association of America, 1999.

Stewart, Andrew. "The Canon of Polykleitos: A Question of Evidence." *American Journal of Archaeology* 98 (1978): 122–31.

Taub, Liba. *Aetna and the Moon: Explaining Nature in Ancient Greece and Rome.* Corvallis: Oregon State University Press, 2008.

————. *Ancient Meteorology.* New York: Routledge, 2003.

————. *Ptolemy's Universe: The Natural, Philosophical, and Ethical Foundations of Ptolemy's Astronomy.* Chicago: Open Court, 1993.

Teresi, Dick. *Lost Discoveries: The Ancient Roots of Modern Science—from the Babylonians to the Maya.* New York: Simon & Schuster, 2002.

Thomson, J. Oliver. *History of Ancient Geography.* New York: Biblo and Tannen, 1965.

Thorndike, Lynn. *A History of Magic and Experimental Science during the First Thirteen Centuries of Our Era.* 2 vols. New York: Columbia University Press, 1923.

Tobin, Richard. "The Canon of Polykleitos." *American Journal of Archaeology* 79 (1975): 307–21.

Tozer, H. F. *A History of Ancient Geography.* 2nd ed. Notes by Max Cary. New York: Biblo and Tannen, 1971 (1897).

Tuplin, C. J., and T. E. Rihll. *Science and Mathematics in Ancient Greek Culture.* New York: Oxford University Press, 2002.

Turner, Howard R. *Science in Medieval Islam.* Austin: University of Texas Press, 2007.

Vrettos, Theodore. *Alexandria: City of the Western Mind.* New York: Free Press, 2001.

Wickkiser, Bronwen L. *Asklepios, Medicine, and the Politics of Healing in Fifth-Century Greece.* Baltimore: Johns Hopkins University Press, 2009.

Woodcroft, Bennett, ed. *The Pneumatics of Hero of Alexandria.* Trans. Joseph Gouge Greenwood. London: Taylor, Walton, and Maberly, 1851.

INDEX

277